T0139311

MOLECULAR BIOLOGY
INTELLIGENCE
UNIT

Behavioral and Morphological Asymmetries in Vertebrates

Yegor B. Malashichev, Ph.D.
Department of Vertebrate Zoology
St. Petersburg State University
St. Petersburg, Russia

A. Wallace Deckel, Ph.D.
Department of Psychiatry and Neuroscience
University of Connecticut Health Center
Farmington, Connecticut, U.S.A.

LANDES BIOSCIENCE
GEORGETOWN, TEXAS
U.S.A.

EUREKAH.COM
GEORGETOWN, TEXAS
U.S.A.

BEHAVIORAL AND MORPHOLOGICAL ASYMMETRIES IN VERTEBRATES

Molecular Biology Intelligence Unit

Landes Bioscience
Eurekah.com

Please address all inquiries to the Publishers:
Landes Bioscience / Eurekah.com, 810 South Church Street, Georgetown, Texas 78626, U.S.A.
Phone: 512/ 863 7762; Fax: 512/ 863 0081
www.eurekah.com
www.landesbioscience.com

ISBN: 1-58706-105-8

Library of Congress Cataloging-in-Publication Data

Behavioral and morphological asymmetries in vertebrates / [edited by]
  Yegor B. Malashichev, A. Wallace Deckel.
    p. ; cm. -- (Molecular biology intelligence unit)
  ISBN 1-58706-105-8
  1. Developmental neurobiology. 2. Cerebral dominance. 3. Laterality.
  4. Vertebrates--Physiology. I. Malashichev, Yegor B. II. Deckel, A.
  Wallace. III. Series: Molecular biology intelligence unit (Unnumbered)
  [DNLM: 1. Behavior, Animal. 2. Vertebrates--physiology. 3. Body
  Patterning. 4. Laterality.
  QL 751 B4185 2006]
  QP363.5.B445 2006
  612.8--dc22

                                        2006021711

About the Editors...

YEGOR MALASHICHEV is Assistant Professor for Zoology at the Department of Vertebrate Zoology in St. Petersburg State University, Russia, and currently is an Alexander von Humboldt Fellow in the Institute of Anatomy and Cell Biology in Freiburg University, Germany. Main research interests include skeleton development and evolution as well as developmental and evolutionary aspects of vertebrate lateralization. He is the Organiser of the series of symposia on Behavioural and Morphological Asymmetries (first in December 2002 in Bentota, Sri Lanka; second in September 2004 in St. Petersburg, Russia).

A. WALLACE DECKEL is a Professor at the University of Connecticut Health Center in Farmington, Connecticut, where he works as a senior Neuropsychologist in the Department of Psychiatry. He has published widely in the area of cerebral specialization. This work has included the study of lateralized aggression in the lizard *Anolis carolinesis* and monoaminergic control of ethanol consumption in the rodent. In human studies, Dr. Deckel has published on lateralized differences in cognition in populations "at risk" for alcoholism and on asymmetrical brain changes in Huntington's disease patients.

CONTENTS

Section II: Eye Use and Cerebral Lateralization

Section III: Vertebrate Studies of Physiological Asymmetries— Perspectives from the West and the East

Section IV: Novel Concepts in Human Studies of Asymmetrical Functions

EDITORS

Yegor B. Malashichev
Department of Vertebrate Zoology
St. Petersburg State University
St. Petersburg, Russia
Email: malashichev@gmail.com
Chapter 4

A. Wallace Deckel
Department of Psychiatry and Neuroscience
University of Connecticut Health Center
Farmington, Connecticut, U.S.A.
Email: Deckel@psychiatry.uchc.edu

CONTRIBUTORS

Valery V. Abramov
Research Institute of Clinical
    Immunology
Siberian Branch of Russian Academy
    of Medical Sciences
Novosibirsk, Russia
Email: valery_abramov@mail.ru
Chapter 12

Yurii G. Balashov
Pavlov Institute of Physiology of Russian
    Academy of Sciences
St. Petersburg, Russia
Chapter 13

Lev V. Beloussov
Department of Embryology
Laboratory of Developmental Biophysics
Moscow State University
Moscow, Russia
Email: lbelous@soil.msu.ru
Chapter 1

Marina P. Chernisheva
Department of General Physiology
St. Petersburg State University
St. Petersburg, Russia
Email: mp_chern@mail.ru
Chapter 14

Irina A. Gontova*
Research Institute of Clinical
    Immunology
Siberian Branch of Russian Academy
    of Medical Sciences
Novosibirsk, Russia
(*since deceased)
Chapter 12

Gisela Kaplan
Centre for Neuroscience
    and Animal Behavior
University of New England
Armidale, Australia
Chapter 5

Elvina P. Kokorina
Department of Physiology
    and Biochemistry of Lactation
All-Russia Research Institute for Farm
    Animal Genetics and Breeding
St.Petersburg-Pushkin, Russia
Chapter 13

Vladimir A. Kozlov
Research Institute of Clinical
   Immunology
Siberian Branch of Russian Academy
   of Medical Sciences
Novosibirsk, Russia
Chapter 12

Dmitri A. Kulagin
Department of Physiology
   and Biochemistry of Lactation
All-Russia Research Institute for Farm
   Animal Genetics and Breeding
St.Petersburg-Pushkin, Russia
Chapter 13

Vitaly P. Leutin
Novosibirsk State Pedagogical University
Novosibirsk, Russia
Chapter 11

Martina Manns
Department of Biopsychology
Institute of Cognitive Neuroscience
Ruhr-University Bochum
Bochum, Germany
Email:
   Martina.Manns@ruhr-uni-bochum.de
Chapter 2

Elena I. Nikolaeva
Herzen State University
St. Petersburg, Russia
Email: klemtina@yandex.ru
Chapter 11

Helmut Prior
Allgemeine Psychologie I
Goethe-Universität Frankfurt am Main
Frankfurt am Main, Germany
Email: Helmut.Prior@gmx.de
Chapter 7

Alexandra Proshchina
Institute of Human Morphology RAMS
Moscow, Russia
Email: proschina@mtu-net.ru
Chapter 3

Lucia Regolin
Department of General Psychology
University of Padua
Padova, Italy
Email: lucia.regolin@psico.unipd.it
Chapter 6

Larissa Yu Rizhova
Department of Physiology
   and Biochemistry of Lactation
All-Russia Research Institute for Farm
   Animal Genetics and Breeding
St.Petersburg-Pushkin, Russia
Email: breusch@pc.dk
Chapter 13

Andrew Robins
Independent Researcher
Australia
Email: arobins@operamail.com
Chapter 8

Lesley Rogers
Centre for Neuroscience
   and Animal Behavior
University of New England
Armidale, Australia
Email: lrogers@une.edu.au
Chapters 5, 10

Sergey Saveliev
Institute of Human Morphology RAMS
Moscow, Russia
Chapter 3

Giorgio Vallortigara
Department of Psychology
B.R.A.I.N. Centre for Neuroscience
University of Trieste
Trieste, Italy
Email: vallorti@univ.trieste.it
Chapter 9

Elena Vershinina
Pavlov Institute of Physiology of Russian
   Academy of Sciences
St. Petersburg, Russia
Chapter 13

PREFACE

This volume grew out of the 2nd International Symposium on Behavioral and Morphological Asymmetries, which took place in St. Petersburg (Russia) in September 2004 at the St. Petersburg State University under the patronage of the St. Petersburg Society of Naturalists. The Symposium is the descendant of a satellite event with a similar name of the 4th World Congress of Herpetology (December, 2001, Bentota, Sri Lanka). While the 1st Symposium (see special issue number 3 for 2002 of the journal, *Laterality*) covered only asymmetries observed in amphibians and reptiles, the second one had a broader scope. Three years passed since the Sri Lanka meeting and there was sustained and increasing interest in vertebrate lateralization in the scientific community, especially in lower vertebrates, or at least, in nonmammalian models. This supported not only by the collection of talks at the Symposium, but also by current publications in international periodicals. Talks here were substantially biased towards the lower vertebrates and birds, while reptiles remained to be studied in more detail.

Two important rationales were considered for the Symposium and the volume, which you have in hand. The first was to bring together topics and specialists representing different branches of the relatively broad field of research of animal asymmetries. The contributions focused on three main subjects: (1) development of structural and functional asymmetries constituted; (2) evolution and adaptation; and (3) function. Aiming for a broader range of topics, the Symposium may still show the current perspective. The increasing number of contributors (twice as many as at the Sri Lanka meeting) give at least a hope that it was indeed so. We, however, further invited authors, who although not present at the meeting itself, nevertheless could contribute to the book to finalize its shape. The other purpose of this volume is to expose Western scientists to Eastern thoughts regarding laterality, and vice versa. We aimed also to help Russian scientists with limited resources and access to the international journals the chance to publish in the Western literature. It seemed to us that this is a fine and perfectly acceptable approach, which on the other hand explains some of the unevenness in the quality and the style of the different manuscripts.

Compared to the program of the Symposium, the structure of the volume is different. This change was done in order to present the Chapters in the most suitable format for the reader and combine them in blocks of interest. Thus the first section comprises Chapters covering the four most intriguing and contradictory questions of development of neurobehavioral asymmetries, respectively—symmetry break in early development and fulfillment of the P. Currie principles (Ch. 1 by L. Beloussov), epigenetics (light exposure in the developing chick: Ch. 2 by M. Manns), and space flight condition (Ch. 3 by A. Proshchina and S. Saveliev) influences on asymmetry formation, as well as the developmental and evolutionary relationship of the neurobehavioral and visceral asymmetries (Ch. 4 by Y. Malashichev). The first and the fourth Chapters also cover comprehensively the issues of morphological and physiological asymmetries found in invertebrate animals.

The second section of this volume is devoted to an intensively explored and therefore rather specialized and very well documented part in the field of animal lateralization, i.e., eye-and-hemisphere specialization, mostly based on work with the avian models. Thus three out of four Chapters in the section are on bird lateralization. Chapter 5 by L. Rogers and G. Kaplan on lateralization of reactions to predators and food in Australian magpies observed in the wild is followed by a review of lateralization in more sophisticated behaviors in domestic chick studied in laboratory (Ch. 6 by L. Regolin), and another review of the intriguing phenomenon of lateralization in magnetic orientation in several species of migrating birds (Ch. 7 by H. Prior). An analytical review on visual lateralization in anuran amphibians closes the section (Ch. 8 by A. Robins).

In the third section we put together contrasting views from the Western and Eastern perspectives on the functioning of the asymmetric brain, its role in adaptation, evolution and species survival (compare for instance Chapters 9 by G. Vallortigara and 10 by L. Rogers and Chapter 11 by E. Nikolaeva and V. Leutin). While the first of this set comprehensively covers more than a 15 year history of chick lateralization studies by a group of Italian neuropsychologists, the second integrates the results from avian and primate models, the third covers (among the other questions) diverse and remote human populations. The methods, approaches and conclusions differ, which makes these papers more interesting for the readers. Chapter 12 by V. Abramov and co-authors describes an experimental attempt of an integrative approach to the study of physiological asymmetries and describes concordant changes in several organ systems. Chapter 13 by L. Rizhova and others seems also to supplement well the current studies of animal visual lateralization (see second section of this volume) in that it shows directional influences of the environment on the physiological status of the animal through the lateralized functioning of the nervous system. The last, but not least, Chapter 14 by M. Chernisheva, which was originally thought to be included in the third section, now stands alone in section four as it presents a hypothesis that the formation of interhemispheric asymmetry is under the influence of regulatory mechanisms of the brain's energy homeostasis—a concept which has never been exposed to the Western reader.

Taken together, these fourteen Chapters, we believe, display a variety of the most interesting and intriguing topics within the broad field of animal lateralization, showing the perspectives of its developments. Far from complete, the volume nevertheless is a state-of-the-art book, which complements a bulk of recent literature on genetics and developmental studies of asymmetries of the heart and other inner organs, interhemispheric specialization in human subjects, and fluctuating morphological asymmetry in animals.

We hope the reader will enjoy the book and accept our approaches and topic choices. To this we should only add our sincere thanks to all reviewers and the professional team of Landes Bioscience, whose efforts helped us substantially to make the volume interesting and well organized, as well as to many colleagues in St. Petersburg State University and St. Petersburg Society of Naturalists for their help in organization of the meeting itself. The editorial work was partially supported by Alexander von Humboldt Stiftung, granting excellent working opportunities for Y. Malashichev in the University of Freiburg.

Yegor B. Malashichev, Ph.D.
A. Wallace Deckel, Ph.D.

SECTION I
Development of Behavioral and Brain Asymmetries

CHAPTER 1

Symmetry Breaks in Early Development of Multicellular Organisms:
Instabilities and Morphomechanics

Lev V. Beloussov*

Abstract

The development of all animal embryos is accompanied by several symmetry breaks which transform a spherically symmetric oocyte into a body, characterized by polarity, dorso-ventrality, left-right and translational dissymmetry. We explore whether these symmetry breaks obey a classical Curie principle, demanding the existence of external "dissymmetrizers", and show that in many cases the symmetry breaks occur "spontaneously", i.e., as a result of instability of a more symmetric shape. Therefore the symmetry breaks should be considered as the consequences of morphogenetic processes rather than the elements of an externally superimposed prepattern. A model is proposed from which predictions can be made about spontaneous symmetry breaks from general morphomechanical laws. We discuss also the conditions permitting a left-right dissymmetry to emerge from a supramolecular to a morphological level.

Introduction

Virtually all Metazoa embryos undergo during early development several steps of reduction of symmetry (symmetry breaks), especially if taking as a criteria of a symmetry order so called "color" (qualitative) characteristics, rather than only geometrical ones. Thus, an immature oocyte, at least prior to the start of vitellogenesis, can be qualified, regardless of its geometrical shape, as a body of a highest possible symmetry (that of a sphere: $\infty / \infty \, m$) because any one of its axes can become the main polar (animo-vegetal) axis of a mature egg. The axis position becomes ultimately determined as the extrusion of polar bodies (more precisely, the second polar body[1]). From this moment on the symmetry of egg and early embryo is reduced to $\infty \cdot m$. The next symmetry break is associated with the establishment of dorso-ventrality, when the symmetry order reduces to $1 \cdot m$ (Fig. 1). In not all Metazoa is this latter symmetry break well expressed. Some called Radiata (to these belong Coelenterata, adult Echinodermata and several other groups of invertebrates), look as retaining, at least in some periods of their life cycle, a perfect radial symmetry, although of a diminished order ($n \cdot m$, rather than $\infty \cdot m$). More careful investigations show however that some elements of dorso-ventrality are acquired by these organisms as well, although in a more or less hidden form.[2] Others, called Bilateria, obtain $1 \cdot m$ symmetry in early development and do not lose it during the entire life cycle. A number of further morphogenetic events can be qualified as

*Lev V. Beloussov—Laboratory of Developmental Biophysics, Department of Embryology, Faculty of Biology, Moscow State University, Moscow, Russia. Email: lbelous@soil.msu.ru

Behavioral and Morphological Asymmetries in Vertebrates, edited by Yegor B. Malashichev and A. Wallace Deckel. ©2006 Landes Bioscience.

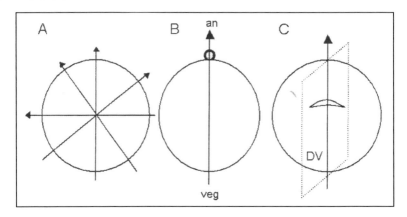

Figure 1. Symmetry breaks in egg cell development. A) An immature oocyte may be potentially polarized along any one of its central axes. B) After polar body extrusion an oocyte acquires a stable polar (animo-vegetal) axis. C) A fertilized egg acquires a dorso-ventral plane (DV) which is marked in amphibian eggs by a cortical crescent-like structure (so called gray crescent).

successive reductions of a translation symmetry order. In Radiata it is a splitting of rotational shapes into an increased number of sectors, or septae (Fig. 2A,B), while in Bilateria it is metamerization, or segmentation of the pre-existing uniform cell condensations (Fig. 2C,D).

Obviously, when ascribing to an embryonic body a certain symmetry, we always make an enormous abstraction by neglecting a lot of structural details which, if taken into consideration, will always reduce symmetry order up to *1*. Such an abstraction means that we always take a certain structural level as a main one, intentionally ignoring (probably, temporarily) all the others. This procedure is absolutely necessary not only for symmetrology, but for all aspects of a general taxonomy and morphology. Meanwhile, it would be instructive to change, from time to time, the structural level to which we focus our attention and to compare the dynamics of the symmetry order changes taking place on the different levels (for more details see ref. 4).

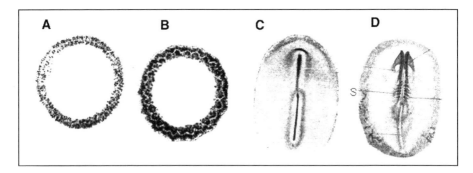

Figure 2. Examples of translational dissymmetrization during advanced development in Radiata (A,B) and Bilateria (C,D). A) Cross-section of a site of tentacles formation in the growing bud of a hydroid polyp *Obelia*, still translationally uniform. B) Same site already split into tentacle rudiments. C) Primitive groove and a so-called head process (a common rudiment of axial organs) in hen's egg blastoderm are still translationally uniform. D) Break of translation symmetry due to splitting of axial mesoderm into several somites (S). A,B) Original. C,D) From Gurwitsch, 1909.[3]

Quite a special kind of symmetry is a left-right (LR-) dissymmetry (absence of a mirror symmetry), which is also called handedness for Bilateria and chirality for Radiata. At the present time it is well established that LR-dissymmetry exists at molecular or, more precisely, supramolecular level, being in most cases associated with the chirality of microtubules, inherited by flagellae.[5-7] As suggested in reference 8, a chirality of spiral cleavage in molluscs is determined by peculiar cholesteric associations of microfilaments in the subcortical layer of eggs, hence by supramolecular structures again. In any case, LR-dissymmetry persists in the molecular structures of living bodies, rather than arising de novo as did the above-mentioned polarity or dorso-ventrality. Correspondingly, the main question associated with LR-dissymmetry is not how it originated, but how can it emerge during development from a supramolecular to a macromorphological level. This is indeed an enormous jump, covering no less than five orders of linear dimensions (roughly from 10^{-8} to 10^{-3} m). For being able to provide such a jump, the elementary molecular events should acquire some kind of a holistic order, or a coherency (see ref. 9 for discussion on this very important property).

Do Symmetry Breaks in Early Development Obey Curie's Principle?

As seen from this brief exposition, a symmetrology of the developing embryos has several different aspects, closely connected with such fundamental biological problems as a collective order (coherency) and interlevel relations; of these we will discuss only a mostly straightforward one: what are the factors providing symmetry breaks in early development and how are these factors related to other morphogenetic phenomena? It is necessary to start this discussion by referring to the famous principle of the French physicist Pierre Curie put forward more than a century ago:

"When the given causes generate the given consequences, the elements of symmetry of the causes should manifest themselves in their consequences. If the events show a certain dissymmetry, the same dissymmetry should be revealed in their causes." [10]

For our purposes, the second phrase is mostly important. Briefly speaking, it forbids what may be called a "spontaneous" break of symmetry: according to Curie's principle, such a break cannot occur within a symmetric body, settled into a symmetric environment. A dissymmetry can only be transferred from one body to another.

For a long time, Curie's principle was regarded as one of the milestones of a classical physics. Interestingly, the developmental biologists of the past generations intuitively followed this principle (although they never referred to Curie), stubbornly searching each next case for its own "external dissymmetrizer". Meanwhile, in recent time the situation in physics, as related to Curie's principle, became more dubious. This was associated with the emergence of a self-organization theory (SOT) pointing to a vast spread of unstable states both in living and nonliving nature (among an enormous number of works about SOT the following ones may be recommended for biologists: refs. 11-13). According to SOT, if a body is in an unstable state, it may reduce its symmetry order under infinitesimal perturbations. That means that although Curie's principle formally retains its validity, for unstable states it becomes, so to say, nonconstructive: if external "dissymmetrizers" are very weak, they cannot be discriminated from "noise", that is, from a continuous spectrum of external perturbations to which all the living systems are permanently exposed. Therefore, we may conclude the following: if one cannot find any "strong enough" (macroscopic) external "dissymmetrizer" for a given symmetry break, the existence of such a break indicates that a system is unstable. So, nonobedience to Curie's principle is itself of substantial heuristic value, in so far as instability is associated with a number of other properties important for morphogenesis. Correspondingly, our task will be to review whether the above-described symmetry breaks are really associated with the presence of any definite external dissymmetrizers.

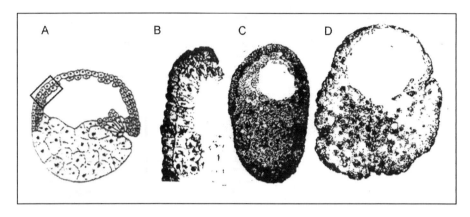

Figure 3. Spontaneous formation of a polarized amphiblastula-like body out of a small piece of a blastocoel roof of *Xenopus laevis* early gastrula embryo. A) Scheme of operation. Framed is an extirpated region. B) A tissue piece immediately after extirpation. C,D) 5 and 18 h after isolation. Reproduced with permission from Beloussov and Petrov, 1983.[17]

Polarization of Eggs (∞ / $\infty \cdot m \to \infty \cdot m$)

A classical object for studying this process are the eggs of brown algae (*Fucus, Pelvetia* and others). Initially it was found that they can be effectively polarized by a directed light: their future rhizoid poles are always oriented opposite to light. So it looked as if these objects strictly obeyed Curie's principle. A final conclusion from these studies was meanwhile quite unexpected: the eggs were found to be polarized and developed quite normally, even if they "are isolated from other cells and from diffusion barriers, kept in dark, and exposed to a gravitational field of only 1g—a vector which does not polarize them".[14] Therefore, their nonpolarized state should be qualified as unstable.

Now what about the animals' eggs? They are usually oriented within a gonad in such a way that their future vegetal pole is in contact with the gonad wall while the animal pole points to a free space. It was widely believed that this orientation plays a role of an external dissymmetrizer (e.g., refs. 15,16). Truly, this dissymmetrizer can be overridden, say, by a centrifugal force, shifting the polar body spindle into an abnormal position,[1] but in this case again a substantial force is involved. Thus, at first glance the animal eggs' polarization fully obeys Curie's principle.

But let us now move towards more advanced developmental stages where this polarization becomes morphologically expressed. For amphibian embryos, this is a blastula stage (so-called amphiblastula) characterized by a thin roof formed by relatively small cells, a thick bottom consisting of large yolk-rich cells and an eccentrically shifted blastocoel. Normally, the roof develops from the animal egg's hemisphere while the bottom from the vegetal one, so that egg polarity is directly transmitted to that of an amphiblastula and seems to be an obligatory prerequisite of a latter. However, a very simple experiment[17] shows that this is not the case: if dissecting from a blastocoel roof of an early gastrula stage embryo just a small piece of tissue, within several hours it becomes spontaneously transformed into a very much diminished, but precise model of an amphiblastula with a thin roof and a thick bottom (Fig. 3). This body polarity has been created de novo in an absolutely spontaneous way and without any relation to the previously established animo-vegetal axis. Again, we have to conclude, that a nonpolar (spheroid) shape of a blastula is unstable and has an intrinsic tendency to pass towards a polar one even in the absence of any detectable external dissymmetrizer, although during normal development such a dissymmetrizer really exists. This situation is known in SOT. If taking, for

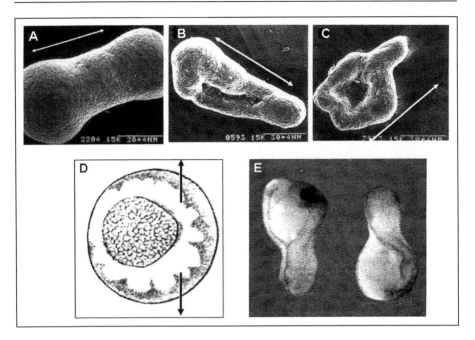

Figure 4. Symmetry breaks after stretching *X. laevis* early gastrula tissues. A-C) Results of stretching of the sandwiches of ventral ectoderm in the directions shown by white arrows. Only in A an imposed $2 \cdot m$ symmetry is preserved, while in B and C it is reduced up to $1 \cdot m$. D) A scheme of a transversal stretching of a suprablastoporall area. E) 24 h results. The stretched embryos acquire $1 \cdot m$ symmetry in the transversal direction instead of a normal $2 \cdot m$ one. Modified from Beloussov et al, 2000.[18]

example, one and the same system possessing both linear, quadratic and cubic negative feedback (for details see ref. 4), its symmetric solution can be either absolutely stable, metastable (demanding finite external perturbations) or unstable depending upon the parameter values. Here we are confronted for the first time with a power of a so-called parametric regulation (a regulation via the parameter values spread homogeneously throughout the entire system). Later on we will return to this topic.

What happens in amphibian development under experimental conditions takes place in the normal development of cnidarians. In cases of anarchic cleavage and so-called multipolar (rather than uni- or bi-polar) migration of endodermal cells, egg polarity is completely "forgotten" during subsequent development so that the polar axis of larva (as well as its rudimentary dorso-ventrality) is established de novo. At a certain stage the embryo looks as an agglomeration of several multicellular toroidal bodies, each gradually transforming into a sphere (by closing its central hole). The last turn determines the position of the posterior pole of a larva.[2] Here we are confronted by a peculiar phenomenon of a spatial symmetry break based upon a temporal pattern. Later on we will mention other such examples.

Morphological polarity can arise de novo also as a result of stretching embryonic tissues either onto an elastic adhesive substrate or by two needles (ref. 18 and in preparation). Obviously, a symmetric stretch itself can produce a body of no less than bi-axial symmetry order $(2 \cdot m)$. Meanwhile, $2 \cdot m$ symmetry of stretched samples is a rare event (Fig. 4A), while a polar symmetry $(1 \cdot m)$ is much more ubiquitous (Fig. 4B-D). This leads to a similar conclusion: polarized shapes are more stable than bipolar ones. Same conclusion comes from SOT.

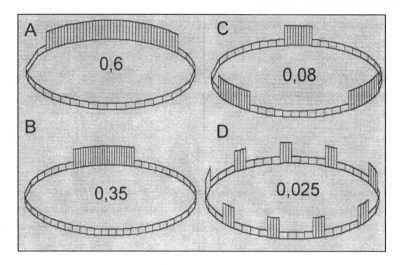

Figure 5. Generation of one-domain and multi-domain (metameric) patterns under the different values of a global threshold (GT) parameter (shown by figures). A diminishment of GT value decreases a single domain length and increases the domains' number (from Beloussov and Grabovsky, 2005).[25]

Establishment of a Dorso-Ventrality ($\infty \cdot m \to 1 \cdot m$)

The situation is similar to that described above. Although in many cases of normal development one can point to some definite external dissymmetrizers (among them, a point of sperm entry into an egg cell, which determines in many species a dorso-ventral meridian), dorso-ventrality can be also established without such agents. This was shown, in particular, by an elegant experiment inserting a sperm precisely into the animal pole of amphibian egg (in which position it cannot select a single dorso-ventral meridian out of infinite meridians). The resulting embryos revealed nevertheless a perfect dorso-ventrality.[19] The same is observed in the cases of parthenogenesis, when no sperm at all is present. Dorso-ventrality seems to be established more or less spontaneously in the eggs of bony fishes.[20] Even in many Spiralia, in spite of their highly determinative cleavage, a ventral or a dorsal fate of the vegetal blastomeres is nonpredetermined; it is a matter of a random choice.[21]

Translational Dissymmetrization ($\infty \to n$)

Both in Radiata and Bilateria this kind of symmetry break definitely lack any external dissymmetrizers. Such a conclusion can be made a priori because no blueprints for these kinds of symmetry can be found at all in the external environment. Therefore, in these cases the assumption of instability is unavoidable. The question is only when and by what means the instability is established. Concerning segmentation of mesoderm in vertebrate embryos, several cogent hypotheses (although if not applicable to all the cases) postulate the transition of a temporal (oscillatory) pattern of gene expression into a spatial one.[22,23] Such a temporal-spatial symmetry transition does not exclude the participation of purely spatial (mechanical) instabilities in establishing segmental patterns.[24]

Deriving Symmetry Breaks from a Morphomechanical Model

So we conclude that the symmetry breaks in the developing embryos can be the result of instability of more symmetric shapes (or temporal patterns). Accordingly, these breaks should be studied with the use of SOT, that is, by constructing models implying nonlinear feedback between the dynamic components of a given system. In modern SOT several families of such models are developed, based either upon chemokinetical or mechanical principles, or both together. Among these, we intend to select that one, which may be applied not only to some individual symmetry breaks, but also to the morphogenetic processes occurring in between the breaks. In our research group we are developing a morphomechanical model of such a kind.[4] Its main idea, which can be only briefly described here is the existence of nonlinear interactions between the active (that is, generated within a given cell or a tissue piece) and passive (coming from other parts of the same embryo) mechanical stresses of tension or compression. For example, according to the model if a cell or a group of cells within a cell sheet becomes tangentially contracted and hence stretches another part of the same sheet, this latter should actively extend in the same direction, compressing the first part. The evidence in favor of this suggestion have been presented elsewhere.[18] Now our question of the model will be whether such a seemingly monotonous interaction can lead, under certain values of controlling parameters, to the instability of initial shape and hence to a symmetry break.

Let us first use this model for making some qualitative estimations concerning spontaneous formation of an amphiblastula-like body (see Fig. 3). We start from a spherically symmetric blastula with its cavity pressurized by osmotically-driven turgor. Now let us apply to the blastula wall just a small local perturbation making a part of it (A) slightly thinner than at rest. So far as tensile stress is a ratio of a stretching force to the transverse section area to which it is applied, the stress in A region will be the greatest. According to our model, this will lead to the active tangential extension of A-region cells and hence to a further (now active) thinning of this region and the relaxation of the rest of the wall. The cells in the latter (again according to the model) will tend to diminish the surface area by contracting and/or by migrating inside. In such a way any slight difference in the wall thickness will increase. One can see that the instability of a spherically symmetric shape is easily derived from our model.

For a quantitative analysis, we plotted a closed circle of a constant length consisting of a constant number of cells and investigated the consequences of its local perturbation by tangential contraction of a single cell.[25] Due to contraction, the resting part of a circle, according to our model, should at first passively stretch and then actively extend. It turned out that the final results crucially depended upon a stretch threshold required for coming from the passive stretch to the active extension. If this threshold was high enough (that is, a substantial stretch was required for the cells' active extension), a single extended domain of tangentially contracted (and hence perpendicularly elongated) cells is formed (Fig. 5A). By reducing the threshold, a number of such domains increases and their lengths correspondingly decrease (Fig. 5B,C). In any case, the translational symmetry of the initial circle is broken and different radial (or metameric) patterns appear.

As another modification of the same model, we explored the morphogenesis of a similar circle consisting of N kinematically independent elements (not necessarily individual cells) each exerting a tangential pressure upon its two neighbors.[26,27] The pressure was assumed to be resisted by a passive elastic force measured by W parameter $(0 < W < 1)$. We also assume that the pressure impulse starts from a certain element and then rotates around the circle either clockwise or counterclockwise. Therefore, we imply an internal handedness of a circle, so far nonassociated with any macromorphological patterns. Accordingly, such a handedness can imitate a molecular chirality, up to now not manifest macroscopically. Our questions to the model are: (1) does it permit production of any radial *macro*-structures whose angular wavelength is greater than $2\pi/N$? (2) should these macrostructures, if arisen, always become chiral, or under some conditions chirality remains hidden and perfectly radial shapes arise, while under other condition values a chirality comes to a macroscopic level?

Quite unexpectedly the radial structures with a wavelength greater than $2\pi/N$ can be produced only if the pressure is applied with some temporal periodicity (that is, the moments of its application are somehow alternated with the periods of the elastic force domination). Under constant pressure nothing except short-range "rack-wheels" is produced (Fig. 6A). This is another indication of the critical relationship between temporal and spatial symmetry. If meanwhile the pressure is applied in a periodic fashion, the answers to both questions become

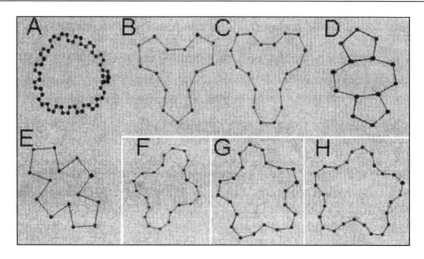

Figure 6. Formative capacities of circular shapes under different N and W values. A) A monotonous "rack-wheel" is produced under the action of a constant (nonperiodic) force at any N and W values. B-H) Radial shapes containing from 3 to 5 lobes, produced under periodic pressure pattern shown in Figure 7 B_1. B) $N = 15$, $W = 0,47$; C) $N = 18$, $W = 0,49$; D) $N = 14$, $W = 0,41$; E) $N = 18$, $W = 0,47$; F) $N = 22$, $W = 0,47$; G) $N = 22$, $W = 0,47$; H) $N = 26$, $W = 0,49$. Note that similar shapes can be reproduced under different combinations of the both parameters (from Beloussov and Grabovsky, 2003b).[27]

positive: a large repertoire of regular macromorphological structures, either radially symmetric or chiral, can be produced within the framework of the model. One of the main lessons from the model was that the resulting patterns crucially depend upon the *combinations* of N and W parameters, rather than their separate values. Figure 6B-H display several absolutely stable radial patterns of the different symmetry orders generated after several dozens of iterations under N and W values shown in the legend. One can see, that very similar (although not completely identical) 3, 4 and 5-fold shapes can be generated by different combinations of parameters. This may be one of the reasons for ubiquitous morphological parallelisms in the animal kingdom.

Meanwhile, a majority of the combinations of parameters, for example intermediate to those shown in Figure 6, produce unstable permanently rotating shapes. In these cases, a superposition of successive iterations perfectly reproduces different chiral patterns. Therefore, under these parameters' values a hidden chirality emerges at a macroscopic level while under the values shown in Figure 6 it remains subtle. In this respect it is of interest to point out that in some evolutionarily ancient groups of Radiata the "canonical" radial forms are "surrounded" by a number of chiral ones, regarded as unsuccessful attempts to reach a stable radial symmetry (Fig. 7).

Conclusions

More than 100 years ago, Pierre Curie claimed:[10] "C'est la dissymmetrie qui cree l'event" ["This is a dissymmetry which creates an event", in the sense: this is a symmetry break which makes a given body distinguishable from its environment]. This aphorism holds true for the developing organisms as well: symmetry breaks are real milestones in their developmental histories. However, the symmetry breaks may be interpreted in quite different ways.

Up to now in embryology textbooks a traditional view is expressed regarding the elements of symmetry (polar axes, dorso-ventral planes and even the elements of a translational symmetry) as something from outside and then becoming the components of a reference system, dictating to the elements of embryonic bodies their final fates. A popular concept of "positional

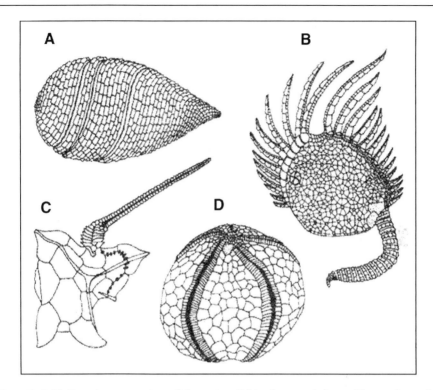

Figure 7. A-D) Several representatives of the ancient Echinodermata. A form with a perfect radial symmetry (D), which is canonical for all the contemporary representatives is surrounded by those with a chiral dissymmetry (A-C), regarded as uncompleted attempts to approach the radial symmetry. A) Helicoplakoidea, B) Paracrinoidea, C) Stylophora, D) Drimitive Echinoidea (from Beloussov and Grabovsky, 2003b, with permission).[27]

information"[28] is an example of such an approach. We oppose to this concept another point of view regarding the symmetry breaks as coming from inherent morphomechanics of an embryonic body itself (although affected by heterogeneities of external environment if the latter are present). By this view the "axes" and "symmetry planes" and, even more, the elements of translational symmetry are the consequences of morphogenesis rather than primary reference systems, which probably do not exist at all or, at least, are not necessary. As we tried to show in this chapter, such a viewpoint is in a better accordance with empirical evidence and permits one to establish some unexpected connections with evolutionary problems. The latter is possible because the parameters used in our model can be, due to their spatial homogeneity, fully open to genetic encoding (fulfilling the condition of a genome equivalency in all the somatic cells).

Being an integral part of morphogenesis, the symmetry breaks in the development of organisms require, by our view, the movements of embryonic parts and perception of the resulting mechanical stresses. These mechano-motile reactions can be considered elementary behavior of an embryo. Correspondingly, the basis for early embryonic symmetry breaks is much the same as for overtly behavioral asymmetries.

Acknowledgements

I am greatly indebted to Dr. Malashichev. This study was supported by the Russian Fund for Basic Researches, grant #05-04-48681.

References

1. Guerrier P. Origine et stabilite de la polarite animale-vegetative chez quelques Spiralia. Ann Embryol et morphogen 1968; 1:119-139.
2. Kraus Ju A, Cherdantzev VG. Experimental study of AP axis formation in the early development of marine hydroid Dynamena pumila. Ontogenez Russ J Devel Biol 2003; 36:365-378.
3. Gurwitsch AG. Atlas and essay of embryology of vertebrates and a man. Prakticheskaya Medizina. St Petersburg (in Russian) 1909.
4. Beloussov LV. The dynamic architecture of a developing organism. Dordrecht, Boston, London: Kluwer Academic Publishers, 1998:238.
5. Cartwright JHE, Piro O, Tuval I. Fluid-dynamical basis of the embryonic development of left-right asymmetry in vertebrates. Proc Natl Acad Sci 2004; 101:7234-7239.
6. McGrath J, Somio S, Makova S, et al. Two populations of node monocilia initiate left-right asymmetry in the mouse. Cell 2003; 114:61-73.
7. Yost L. Vertebrate left-right development. Cell 1995; 82:689-692.
8. Mescheryakov VN. How the genes distinguish right and left? In: Presnov EV, Maresin VM, Ivanov AV, eds. Analytical Aspects of Differentiation Moskva, Nauka 1991:137-166, (in Russian).
9. Ho Mae Wan. Rainbow and the Worm. World Scientific, Singapore: 2003.
10. Curie P. De symmetrie dans les phenomenes physique: Symmetrie des champs electriques et magnetiques. J de Physique Ser 1894; 3:393-427.
11. Prigogine I. From Being to Becoming. New York: WH Freeman and Company, 1980.
12. Nicolis G, Prigogine I. Exploring Complexity. New York: WH Freeman and Co, 1989.
13. Capra F. The Web of Life. New York: Anchor Books, 1996.
14. Jaffe L. Localization in the developing Fucus eggs and the general role of localizing currents. Adv Morphogen 1968; 7:295-328.
15. Child CM. Patterns and problems of development. Chicago: University of Chicago Press, 1941.
16. Raven CP. The cortical and subcortical cytoplasm of the Lymnaea egg. Int Rev Cytol 1970; 28:1-44.
17. Beloussov LV, Petrov KV. Role of cell interactions in the differentiation of the induced tissues of amphibian embryos. Ontogenez Sov J Dev Biol 1983; 14:21-29.
18. Beloussov LV, Louchinskaia NN, Stein AA. Tension-dependent collective cell movements in the early gastrula ectoderm of Xenopus laevis embryos. Dev Genes and Evol 2000; 210:92-104.
19. Nieuwkoop PD. Origin and establishment of an embryonic polar axis in amphibian development. Curr Top Devel Biol 1977; 11:115-117.
20. Cherdantzeva EV, Cherdantzev VG. Determination of a dorso-ventral polarity in the embryos of a fish, Brachydanio rerio (Teleostei). Ontogenez Sov J Devel Biol 1985; 16:270-280.
21. Arnolds W, van den Biggelaar J, Verdonk N. Spatial aspects of cell interactions involved in the determination of dorsoventral polarity in equally cleaving Gastropoda and regulative abilities of their embryos as studied by micromere deletion in Lymnaea and Patella. W. Roux's Arch Dev Biol 1983; 192:75-85.
22. Cooke J, Zeeman EC. A clock and wavefront model for the control of the numbers of repeated structures during animal development. J Theor Biol 1976; 58:455-476.
23. Pourquie O, Dale K, Dubrille J et al. A molecular clock linked to vertebrate segmentation. In: Sanders EJ, Lash JW, Ordahl Ch P, eds. The Origin and Fate of Somites. Amsterdam: IOS Press, 2001:64-70.
24. Beloussov LV. Somitogenesis in vertebrate embryos as a robust macromorphological process. In: Sanders EJ, Lash JW, Ordahl Ch P, eds. The Origin and Fate of Somites. Amsterdam: IOS Press, 2001:97-106.
25. Beloussov LV, Grabovsky VI. A common biomechanical model for the formation of stationary cell domains and propagating waves in the developing organisms. Comput Methods Biomech Biomed Engin 2005; in press.
26. Beloussov LV, Grabovsky VI. A geometro-mechanical model for pulsatile morphogenesis. Comput Methods Biomech Biomed Engin 2003a; 6:53-63.
27. Beloussov LV, Grabovsky VI. Formative capacities of mechanically stressed networks: Developmental and evolutionary implications. Riv Biol 2003b; 96:190-197.
28. Wolpert L. One hundred years of positional information. Trends Genet 1996; 12:359-364.

CHAPTER 2

The Epigenetic Control of Asymmetry Formation:
Lessons from the Avian Visual System

Martina Manns*

Abstract

Although lateralization is a core feature of information processing of vertebrate brains, there is no model which can explain how ontogenetic mechanisms lead to an adult asymmetric functional architecture. While the very early appearance of embryonic asymmetries and the heritability of specific lateralization patterns suggest a genetic foundation, a high degree of plasticity highlights the critical role of environmental factors. The avian visual system demonstrates that the formation of neuronal asymmetries can be caused by sensory stimulation that is asymmetrically experienced. Monocular deprivation or intraocular applications of tetrodotoxin or BDNF suggest that lateralization develops via activity-dependent differentiation of brain circuits. A brief period of visual asymmetry in prehatch birds, resulting from a genetically determined head turning bias, triggers asymmetric differentiation processes in both hemispheres which gain significance during posthatch maturation. During this time, functional dominance of the right eye/left hemisphere for visual feature analysis develops, and morphological asymmetries in the tectofugal pathway differentiate into an adult phenotype.

In sum, asymmetry formation in pigeons can be used as a general model to examine how biased peripheral stimulation establishes cerebral lateralization. It can explain how both epigenetic influences and genetically determined left-right differences contribute to the development of laterality.

Lateralization Is a Core Feature of Information Processing in the Vertebrate Brain

As we all know from common experience, humans prefer one—mostly the right—hand for unimanual manipulations. Less well known is the fact that several other cortical functions are also lateralized. For example, language is mainly processed in the left hemisphere while spatial skills or emotional behavior are generally under control of the right hemisphere. Some of these functional asymmetries are associated with left-right differences of gross anatomical landmarks and/or architectonic cortical subdivisions.

*Martina Manns—Department of Biopsychology, Institute of Cognitive Neuroscience, Faculty of Psychology, Ruhr-University Bochum, 44780 Bochum, Germany. Email: Martina.Manns@ruhr-uni-bochum.de

Behavioral and Morphological Asymmetries in Vertebrates, edited by Yegor B. Malashichev and A. Wallace Deckel. ©2006 Landes Bioscience.

Cerebral lateralization originally was considered to be a unique characteristic of the human brain, and it is likely that evolutionary pressures associated with upright body position, tool use and speech may have selected out asymmetries that favor survival. Recent research demonstrates the existence of lateralization in all vertebrate species, placing lateralization as an ancient feature of the vertebrate brain.[1,2] Although animal models provide the opportunity to examine phylogenetic and developmental foundations of cerebral lateralization, the functional and ontogenetic interplay between neuronal substrate and behavioral lateralization is still an unsolved problem. This ambiguity results at least partly from the uncertainty regarding the relative contribution of genetically versus environmentally determined lateralization.

Unsolved Riddle: The Ontogenetic Foundations of Cerebral Lateralization

The presence of a population bias for cerebral asymmetries like handedness or speech processing in humans provoked genetic models to explain the ontogenetic foundations of lateralization.[3] In humans, twin studies support a genetic basis of cortical volume and handedness.[4,5] In fish, data supports the role of genetic factors as the source of a turning bias.[6,7]

The neuronal mechanisms that mediate genetic regulation of asymmetrical brain development are still unclear, but asymmetry formation within the dorsal diencephalon of vertebrates provides some hints.[8,9] The alignment of anatomical asymmetries in the epithalamus is controlled by the Nodal signaling pathway, a gene cascade that is also involved in biasing laterality of the visceral organs.[10,11]

Additionally, early embryonic emergence of behavioral and morphological asymmetries may have a genetic basis. Human fetuses in uterus exhibit lateralized motor behavior,[12-14] display a functional hemispheric asymmetry in auditory evoked cortical activity,[15] and develop asymmetry of the planum temporale, which is regarded as the anatomical basis for lateralized language dominance.[16,17]

Since anatomical left-right differences represent the structural basis for cerebral lateralization they should precede the appearance of behavioral lateralization. Similarly, the direction of anatomical left-right differences should be correlated with functional ones. Such a relationship is found between right-handers and their language dominant left-hemisphere, where a leftward asymmetry in planum temporale exists. But this asymmetry is less pronounced in sinistrals.[5,18] Moreover, pre- and postnatal events can affect asymmetry during development of the planum temporale and disrupt twin concordance.[19,20] This plasticity indicates the critical role of environmental factors. At least in some systems, environmental influences are actually essential for the establishment of cerebral lateralization. The interplay between gene-dependent prespecifications and epigenetic control is exemplified in the adoption of face expertise in human brains. Visual input is necessary to gain face recognition competence but affects only the right hemisphere suggesting that this brain side is predetermined to achieve face recognition competence.[21]

Genetic models cannot explain such plasticity in structure-function bias because the underlining neurobiology that controls this process is unknown. Here, we propose that lateralized environmental experiences during embryonic[22] or postnatal[23] development are crucially involved in the establishment of stable cerebral lateralization patterns.

Although cerebral asymmetries are assumed to control handedness, asymmetric motor behavior arises earlier than structural left-right differences within the cerebral cortex. Asymmetry of the planum temporale develops during the third gestational trimester,[16,17] but human embryos perform more arm movements with their right arm and exhibit a preference for sucking their right thumb from the first trimester gestation onwards.[12-14] The early appearance of lateralized motor behavior suggests a muscular or spinal control because a functional corticospinal tract has not been developed at this timepoint. Such lateralized motor behavior might represent left-right differences in maturational speed e.g., of controlling GABAergic systems within the spinal cord.[13,14] These motor asymmetries are related to postnatal handedness.[24] From the final weeks of gestation to the first six months after birth, neonates develop a preference for turning their head to the right.[25] This positional bias correlates with the preference of the fetus to suck the right thumb.[12] Michel and Harkin[23] propose that the neonatal rightward bias in the direction of head orientation is the starting point in developing a stable hand preference. A turning bias leads to a greater amount of ipsilateral hand and arm movement that, in turn, results in an ipsilateral prehensile grasping preference. This creates differences between the hands in their experience of object manipulation and, hence, a bias in unimanual manipulation. Finally, this unimanual preference may develop into a role-differentiated bimanual manipulation preference.[26] These studies describe developmental steps determining handedness preference by lateralized sensorimotor experience, but they, like earlier cited studies, are not able to clarify the mechanisms by which this occurs.[14]

A deeper understanding of the mechanisms that cause these developmental asymmetries can only be gained by experimental manipulations in animal models. Here, new insight comes from the avian visual system where behavioral lateralization can be associated with morphological left-right differences at the individual as well as the population level. This system suggests that epigenetic factors play a critical role in inducing cerebral lateralization and, in our laboratory, we have undertaken a number of studies to unravel the neuronal mechanisms that cause this asymmetrical development.

Visual Lateralization in the Avian Brain: A Model System for the Neuronal Foundations of Cerebral Asymmetries

An increasing number of cognitive studies shows that the left and right hemisphere of the avian brain analyze different aspects of visual stimuli.[27,28] While the left hemisphere is specialized for detailed visual object analysis allowing rapid categorization of food objects or use of local aspects for spatial encoding,[30-32] the right hemisphere extracts relational or global (geometric) properties of visual stimuli.[33-35] These hemispheric specializations can be easily tested by occluding one eye, as the optic nerves in birds completely cross to the contralateral hemisphere. Specifically, the right eye is connected with the left hemisphere and vice versa while the absence of major commissures in the avian brain allows a restricted information transfer between the two hemispheres.[28,29]

The observed behavioral lateralization is associated with morphological asymmetries in the ascending visual systems.[27,29] Chicks exhibit transient left-right differences in the thalamofugal pathway. This system corresponds to the mammalian geniculostriatal system and transfers retinal information via the contralateral geniculate complex (GLd) bilaterally onto the telencephalic visual Wulst (Fig. 1A).[36] The left GLd gives rise to more projections to the right Wulst than the right GLd to the left Wulst.[37-40] In contrast, in pigeons visual lateralization is related to morphological asymmetries in the tectofugal pathway. This system corresponds to the mammalian extrageniculocortical system which projects via the contralateral mesencephalic optic tectum and the diencephalic nucleus rotundus to the forebrain (Fig. 1B).[36] Apart from tectal[41-43] and rotundal[44] cell size differences, the tectorotundal projection is asymmetrically organized with more fibers ascending from the right tectum to the left rotundus than vice versa.[45] Thus, the left hemisphere receives a stronger bilateral input from both visual hemifields. All recent studies indicate that the control of visuomotor processing is critically dependent on activity of the left hemisphere. This is supported by drastically reduced visual discrimination capabilities after left-sided forebrain lesions or by biochemical manipulations,[27,28] and by studies showing that the left hemisphere regulates bilateral tectofugal processing.[46]

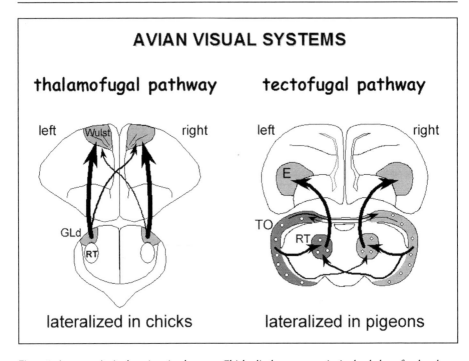

Figure 1. Asymmetries in the avian visual system. Chicks display asymmetries in the thalamofugal pathway with asymmetric projections from the GLd to the Wulst. In pigeons, asymmetries are implemented in the tectofugal pathway with cell size asymmetries in the optic tectum (TO) and nucleus rotundus (RT), and asymmetric ascending and intertectal projections. E: entopallium; GLd: dorsal lateral geniculate complex; RT: nucleus rotundus; TO: optic tectum.

Visual Lateralization Is Actually Induced during Embryonic Development but Consolidated during the Posthatching Phase

The left hemispheric dominance for visual object analysis depends on asymmetric light stimulation during embryonic development.[27,47,48] Prior to hatching, avian embryos keep their head turned to the right such that the right eye is close to the egg shell and the left eye is occluded by the body.[49] Thus, light shining through the translucent shell stimulates the right eye while the left eye is visually deprived. It is likely that this asymmetric position is genetically determined because torsion of the embryo axis is controlled by left/right-specific cascades of gene expressions which also determine heart looping bias.[50] Consequently, incubation of embryos in complete darkness prevents the formation of behavioral as well as anatomical asymmetries.[42,51] In chicks, the normal lateralization pattern can be reversed by occluding the embryo's right eye and exposing its left eye to light for 24 hours.[37,52]

However, chicks hatch as precocial birds with a fully mature visual system, able to forage and follow their mother. In contrast, the altricial pigeon hatches with closed eyes and a highly immature visual system staying three weeks in the nest fed with crop milk by their parents.[53] Therefore, it is possible to alter the final lateralization pattern in pigeons by modulating the visual experience post hatch. Comparable to chicks, occlusion of the right eye in pigeons reverses visual lateralization by inducing a functional dominance of the left eye and by modulating tectofugal left-right differences. Conversely, left eye deprivation enhances right eye dominance.[44,54] Thus, the vulnerable period for the development of visual asymmetries extends into the posthatching period, thus delineating two developmental phases critically involved in the establishment of a lateralized architecture of the pigeon's visual system (Fig. 2).

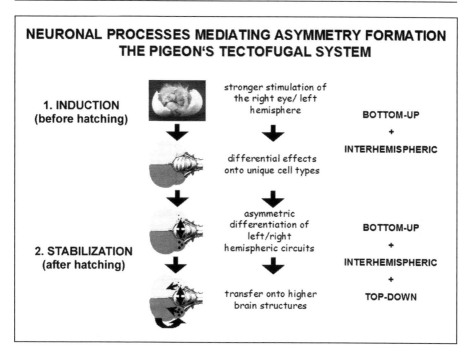

Figure 2. Two phases are critically involved in the development of a lateralized architecture of the pigeon's visual system. In a first step, the visual stimulation of the right eye/left hemisphere during embryonic development induces tectofugal asymmetries by differential effects on unique cell types. Bottom-up and interhemispheric interactions regulate asymmetric differentiation of left and right hemispheric visual circuits. In a second step that occurs post hatch, the induced asymmetries are transferred to higher brain structures by interactions of bottom-up, top-down and interhemispheric projections.

Asymmetries Develop According to Mechanisms Well Known to Be Involved in Ontogenetic Plasticity

In general, neuronal plasticity which exists during the development of the nervous system allows the maturing brain to respond to environmental experiences during critical periods of development.[55] The visual pathways, in particular, have been established as model systems to examine how sensory input controls the activity-dependent development of neurons. As described earlier, the asymmetric head turning in avian embryos causes the amount of incoming light to differ between the left and right eye. In pigeons, this biased photic stimulation causes anatomical left-right differences in the development of the tectofugal pathway. The optic tectum is the first station of the processing stream where morphological asymmetries are visible, with a majority of retinorecipient neurons displaying larger cell bodies in the left tectum.[41] Since the soma size of a neuron is an indicator for the extent of the axo-/dendritic arborization pattern, tectal soma size asymmetries indicate differences in the complexity of left and right tectal circuits.

Since the maturation of the retinotectal pathway is regulated by photic stimulation,[56,57] it is likely that retinal activity differences constitute the first step in the initiation of asymmetric anatomical development. In fact, the transient inhibition of retinal activity by intraocular injections of the sodium channel blocker tetrodotoxin (TTX) leads to a dominance of the ipsilateral nondeprived hemisphere.[58] This activity-dependence suggests that lateralization develops according to mechanisms well known to be involved in activity-dependent maturation of the nervous system.

Brain derived neurotrophic factor (BDNF) may serve as a key player in ontogenetic plasticity.[59-62] Many of the effects of light on the asymmetrical development of the avian visual pathways are mediated by BDNF. For example, light regulates BDNF expression and secretion,[63,64] while BDNF rescues dark rearing effects[65] and affects axo-dendritic dynamics within the retinotectal system.[66-69] Since BDNF is present in the developing retinotectal system of pigeons,[70] it is conceivable that retinal activity differences are mediated by asymmetric BDNF supply. Hence, BDNF application should be able to mimic the effects of a light pulse. This hypothesis was tested by intraocular BDNF injections into the right eye of newly hatched pigeons which were incubated in complete darkness. These injections caused the animals to display a modified adult functional and morphological asymmetry pattern (Manns and Güntürkün in preparation).[71]

BDNF exerts its physiological role by binding to its specific neurotrophic tyrosine kinase (TrkB) receptor. Ligand bound TrkB receptors activate intracellular signaling cascades which, in turn, affect activity and/or differentiation of the responding cells.[72,73] The small intracellular membrane anchored GTPase Ras is a critical molecular switch by which BDNF induces its neurotrophic actions. BDNF/Ras induction thus may signal enhanced cell sizes and axo-dendritic complexity.[74-76] Asymmetric BDNF supply should lead to the asymmetric activation of the TrkB/Ras-signalling cascade. In fact, light incubation during embryonic development leads to a transient inhibition of the TrkB/Ras signalling within the stronger stimulated left optic tectum, but only after hatching.[77] These data verify posthatch consequences of biased embryonic visual experience at the cellular level.

It is very likely that posthatch effects are mediated by inhibitory interactions within the optic tectum.[54] Experience-dependent plasticity in the developing visual cortex is critically regulated by local GABA circuits[78] and GABAergic cells are enlarged in the stronger stimulated left tectum.[43] The critical roles of intratectal inhibitory effects are exemplified at the tectal level (Fig. 3). While the majority of tectal cells display enlarged cell bodies on the left side, supporting a growth promoting effect of light, the efferent cells in the deeper lamina (giving rise to the ascending forebrain projections) are larger in the right tectum.[41] This soma size asymmetry pattern develops within the first week after hatching in response to reduced TrkB/Ras signaling.[77] It is conceivable that the light-dependent stimulation of left-tectal GABAergic input exerts enhanced inhibitory control onto the efferent cells in the left tectum, thus leading to smaller cell bodies and to fewer contralaterally ascending projections arising from the left tectum.[45,54]

Visual Lateralization Results from the Balance of Left- and Right-Hemispheric Differentiation Processes

The consequences of asymmetric light stimulation are not confined to an enhanced trophic support of the left brain. A detailed analysis of light- and dark-incubated animals reveals that light induces a left-hemispheric increase in visuoperceptual skills. Conversely, light simultaneously decreases visuomotor speed within the right hemisphere. Thus, specialized visual circuits are differentially adjusted in both hemispheres.[42] These complex effects may be caused by a differential sensitivity of distinct cell types to retinal input. Evidence for such differential effects can be observed in immunohistochemically characterized tectal cells. Unique cell populations express different calcium-binding proteins like calbindin, calretinin or parvalbumin. In light-incubated animals, parvalbumin-positive cells display smaller cell bodies in the left, light stimulated tectum, indicating a suppressive effect of light on this cell type.[43] Accordingly, intraocular BDNF injection reduces parvalbumin-positive neuron size in both tectal halves. In contrast, calbindin-positive (presumably GABAergic) cells in the retinorecipient tectal laminae are enlarged only in the BDNF-enriched tectum. Calbindin-positive but nonGABAergic neurons in the efferent cell layer are not affected at all.[71]

The complex bihemispheric effects that occur at the behavioral as well as cellular level require control over the balance between left- and right hemispheric circuits. Even subtle retinal

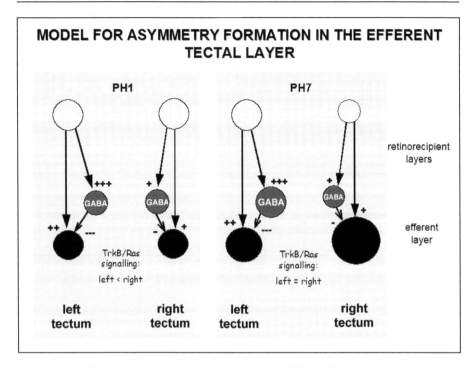

Figure 3. Model for asymmetry formation in the efferent tectal cell layer. Efferent tectal cells receive direct as well as indirect visual input from the outer retinorecipient lamina. Due to the stronger photic stimulation of the right retina, left retinorecipient tectal cells experience greater activation, in turn leading to their enhanced differentiation after hatching (PH1). In particular, maturation and/or synaptic activity of GABAergic interneurons is enhanced within the left tectum. Greater inhibition of efferent cells leads to reduced TrkB/Ras signaling within these cells. This, in turn, develops smaller cell bodies in the left tectum during the first week after hatching (PH7).

modulations are able to interrupt this balance. This conclusion is supported by the transient inhibition of retinal activity with TTX. Dominance of the nonsuppressed ipsilateral eye/contralateral hemisphere can be attributed to a performance increase conveyed by this brain side while performance, when tested with the TTX-injected eyes, does not differ from that of saline-injected controls. The transient silencing of one visual input does not simply suppress the deprived hemisphere but alters the activity balance between the left and right eye, enhancing visuoperceptive skills in the activated hemisphere.[58] A corresponding effect can be observed at the cellular level. Although GABAergic tectal cells are smaller in light-incubated animals compared to dark-incubated ones, this suppressive effect is less pronounced in the stronger stimulated left tectum of light incubated birds which bears larger GABAergic cell bodies.[43]

The necessary integration of activity from the left and the right side might be mediated by inter- and/or intrahemispheric influences.[54] On the one hand, the two tectal hemispheres are connected by mainly inhibitory commissures.[79,80] In pigeons, this interaction is asymmetrically organized with a stronger influence of the left tectum onto the right one than vice versa.[81] After transection of these commissures, lateralization of visually controlled behavior is reversed.[82] On the other hand, tectofugal processing is controlled by afferents from the forebrain and this influence is presumably lateralized. Only left hemispheric lesions of the descending fibre tracts disrupt lateralization,[83] and only the left visual Wulst controls tectofugal processing.[46] This raises the possibility that top-down influences are involved in the final establishment and/ or maintenance of cerebral lateralization.

Synopsis: Lateralization Develops within the Scope of Developmental Plasticity

In sum, in the avian visual system the formation of neuronal asymmetries may occur as a result of an unbalanced sensory stimulation. This, in turn, may lead to a functional lateralization by bottom-up processes that are mediated by mechanisms known to be involved in developmental plasticity.

In avian species, lateralized brain functions start with an asymmetric light trigger. This trigger induces asymmetric differentiation in the tectofugal system. This, in turn, leads to asymmetric interactions with intra- and/or interhemispheric developing circuitries, causing functional lateralization. This process requires a phase during which induced asymmetries must be stabilized and, hence, can be easily modulated. It is likely that inhibitory interactions regulate these processes. Since motor asymmetries in human embryos also precede cerebral lateralization, it is conceivable that human cerebral lateralization develops according to similar developmental principles. For example, the ability of spinally controlled asymmetries to influence the cerebral cortex may represent a human corollary to the avian system.[12-14,25]

The presence of light-independent asymmetries in chicks suggests that some aspects of forebrain control may be independent from visual input. Dark incubated chicks display higher ability to assess and respond to novelty seeing with their left eye,[85] and endogenous asymmetries in receptor binding were demonstrated in telencephalic imprinting areas.[84] However, visual experience can modify these endogenous left-right differences.[84,85] Thus, even when genetically determined asymmetries are present, their direction can be modulated by environmental factors. This supports a role for bottom-up processes in the determination of cerebral lateralization and suggests that genetic factors do not directly lead to a functional lateralization pattern. Rather, it is the interaction between genetically determined and environmental factors that cause asymmetrical regulation by the brain. Inherited asymmetries can provoke left-right differences in a variety of areas, including: (a) the rate in which left-right differences in maturation,[86] growth or susceptibility to epigenetic factors like hormones, sensory input, or motor activity occur, (b) morphogenesis leading to asymmetric body positions or craniofacial asymmetries,[22] each of which results in biased environmental experience, or (c) neuronal substrate like asymmetries in cell number[4] or receptor densities,[84] which cause a differential sensitivity to epigenetic factors.

In summary, lateralization develops as the result of the interplay between genetically determined and epigenetically controlled factors, findings which suggest that lateralization can be explained by mechanisms mediating ontogenetic plasticity. This plasticity may explain why very early peripheral asymmetries or developmental disturbances have such a great impact on the final pattern of cerebral lateralization.

References

1. Bisazza A, Rogers LJ, Vallortigara G. The origins of cerebral asymmetry: A review of evidence of behavioral and brain lateralization in fishes, reptiles and amphibians. Neurosci Biobehav Rev 1998; 22:411-426.
2. Vallortigara G, Rogers LJ, Bisazza A. Possible evolutionary origins of cognitive brain lateralization. Brain Res Rev 1999; 30:164-175.
3. Annett M. Left, right, hand and brain: The right shift theory. London: L Erlbaum Associates publishers, 1985.
4. Rosen GD. Cellular, morphometric, ontogenetic and connectional substrates of anatomical asymmetry. Neurosci Biobehav Rev 1996; 20:607-615.
5. Geschwind DH, Miller BL, DeCarli C et al. Heritability of lobar brain volumes in twins supports genetic models of cerebral laterality and handedness. Proc Natl Acad Sci 2002; 99:3176-3181.
6. Bisazza A, Facchin L, Vallortigara G. Heritability of lateralization in fish: Concordance of right-left asymmetry between parents and offspring. Neuropsychologia 2000; 38:907-912.
7. Bisazza A, Dadda M, Cantalupo C. Further evidence for mirror-reversed laterality in lines of fish selected for leftward or rightward turning when facing a predator model. Behav Brain Res 2005; 156:165-171.

8. Concha ML, Wilson SW. Asymmetry in the epithalamus of vertebrates. J Anat 2001; 199:63-84.
9. Halpern ME, Liang JO, Gamse JT. Leaning to the left: Laterality in the zebrafish forebrain. Trends Neurosci 2003; 26:308-313.
10. Concha ML, Burdine RD, Russell C et al. A nodal signaling pathway regulates the laterality of neuroanatomical asymmetries in the zebrafish forebrain. Neuron 2000; 28:399-409.
11. Concha ML, Russell C, Regan JC et al. Local tissue interactions across the dorsal midline of the forebrain establish CNS laterality. Neuron 2003; 39:423-438.
12. Hepper PG, Shahidullah S, White R. Handedness in the human fetus. Neuropsychologia 1991; 29:1107-1111.
13. Hepper PG, McCartney GR, Shannon EA. Lateralised behavior in first trimester human foetuses. Neuropsychologia 1998; 36:531-534.
14. McCartney G, Hepper P. Development of lateralized behavior in the human fetus from 12 to 27 weeks' gestation. Dev Med Child Neurol 1999; 41:83-86.
15. Schleussner E, Schneider U, Arnscheidt C et al. Prenatal evidence of left-right asymmetries in auditory evoked responses using fetal magnetoencephalography. Early Hum Dev 2004; 7:133-136.
16. Chi JG, Dooling EC, Gilles FH. Left-right asymmetries of the temporal speech areas of the human fetus. Arch Neurol 1977; 34:346-348.
17. Preis S, Jancke L, Schmitz-Hillebrecht J et al. Child age and planum temporale asymmetry. Brain Cogn 1999; 40:441-452.
18. Foundas AL, Leonard CM, Hanna-Pladdy B. Variability in the anatomy of the planum temporale and posterior ascending ramus: Do right- and left handers differ? Brain Lang 2002; 83:403-424.
19. Steinmetz H, Herzog A, Schlaug G et al. Brain asymmetry in monozygotic twins. Cereb Cortex 1995; 5:296-300.
20. Eckert MA, Leonard CM, Molloy EA et al. The epigenesis of planum temporale asymmetry in twins. Cereb Cortex 2002; 12:749-755.
21. Le Grand R, Mondloch CJ, Maurer D et al. Expert face processing requires visual input to the right hemisphere during infancy. Nature Neuroscience 2003; 6:1108-1112.
22. Previc FH. A general theory concerning the prenatal origins of cerebral lateralization in humans. Psychol Rev 1991; 98:299-334.
23. Michel GF, Harkins DA. Postural and lateral asymmetries in the ontogeny of handedness during infancy. Develop Psychobiol 1986; 19:247-258.
24. Hepper PG, Wells DL, Lynch C. Prenatal thumb sucking is related to postnatal handedness. Neuropsychologia 2005; 43:313-315.
25. Ververs IA, de Vries JI, van Geijn HP et al. Prenatal head position from 12-38 weeks. I. Developmental aspects. Early Hum Dev 1994; 39:83-91.
26. Hinojosa T, Sheu C-F, Michel GF. Infant hand-use preferences for grasping objects contributes to the development of a hand-use preference for manipulating objects. Dev Psychobiol 2003; 43:328-334.
27. Rogers LJ. Behavioral, structural and neurochemical asymmetries in the avian brain: A model system for studying visual development and processing. Neurosci Biobehav Rev 1996; 20:487-503.
28. Güntürkün O. Avian visual lateralization: A review. NeuroReport 1997; 6:iii-xi.
29. Güntürkün O. Hemispheric asymmetry in the visual system of birds. In: Hugdahl K, Davidson RJ, eds. Brain Asymmetry. 2nd ed. Cambridge: MIT Press, 2002:3-36.
30. Mench JA, Andrew RJ. Lateralization of food search task in the domestic chick. Behav Neural Biol 1986; 46:107-114.
31. Güntürkün O, Kesch S. Visual lateralization during feeding in pigeons. Behavioral Neurosci 1987; 101:433-435.
32. Valenti A, Sovrano VA, Zucca P et al. Visual lateralisation in quails (Coturnix coturnix). Laterality 2003; 8:67-78.
33. Tommasi L, Vallortigara G. Encoding of geometric and landmark information in the left and right hemispheres of the Avian Brain. Behav Neurosci 2001; 115:602-613.
34. Tommasi L, Vallortigara G. Hemispheric processing of landmark and geometric information in male and female domestic chicks (Gallus gallus). Behav Brain Res 2004; 155:85-96.
35. Vallortigara G, Pagni P, Sovrano VA. Separate geometric and nongeometric modules for spatial reorientation: Evidence from a lopsided animal brain. J Cogn Neurosci 2004; 16:390-400.
36. Güntürkün, O. Sensory physiology: Vision. In: Whittow GC, ed. Sturkie's Avian Physiology. 5th ed. San Diego: Academic Press, 2000:1-19.
37. Rogers LJ, Sink HS. Transient asymmetry in the projections of the rostral thalamus to the visual hyperstriatum of the chicken, and reversal of its direction by light exposure. Exp Brain Res 1988; 70:378-384.

38. Rogers J, Deng C. Light experience and lateralization of the two visual pathways in the chick. Behav Brain Res 1999; 98:1-15.
39. Deng C, Rogers LJ. Prehatch visual experience and lateralization in the visual Wulst of the chick. Behav Brain Res 2002; 134:375-385.
40. Koshiba M, Nakamura S, Deng C et al. Light-dependent development of asymmetry in the ipsilateral and contralateral thalamofugal visual projections of the chick. Neurosci Lett 2003; 336:81-84.
41. Güntürkün O. Morphological asymmetries of the tectum opticum in the pigeon. Exp Brain Res 1997; 116:561-566.
42. Skiba M, Diekamp B, Güntürkün O. Embryonic light stimulation induces different asymmetries in visuoperceptual and visuomotor pathways of pigeons. Behav Brain Res 2002; 134:149-156.
43. Manns M, Güntürkün O. Light experience induces differential asymmetry pattern of GABA- and parvalbumin-positive cells in the pigeon's visual midbrain. J Chem Neuroanat 2003; 25:249-259.
44. Manns M, Güntürkün O. "Natural" and artificial monocular deprivation effects on thalamic soma sizes in pigeons. NeuroReport 1999; 10:3223-3228.
45. Güntürkün O, Hellmann B, Melsbach G et al. Asymmetries of representation in the visual system of pigeons. NeuroReport 1998; 18:4127-4130.
46. Folta K, Diekamp B, Güntürkün O. Asymmetrical modes of visual bottom-up and top-down integration in the thalamic nucleus rotundus of pigeons. J Neurosci 2004; 24:9475-9485.
47. Rogers LJ. Light experience and asymmetry of brain function in chickens. Nature 1982; 297:223-225.
48. Güntürkün O. Ontogeny of visual asymmetry in pigeons. In: Rogers LJ, Andrew RJ, eds. Comparative Vertebrate Lateralization. Cambridge: University Press, 2002:247-273.
49. Kuo ZY. Ontogeny of embryonic behavior in aves. III. The structural and environmental factors in embryonic behavior. J Comp Psychol 1932; 13:245-271.
50. Faisst AM, Alvarez-Bolado G, Treichel D et al. Rotatin is a novel gene required for axial rotation and left-right specification in mouse embryos. Mech Dev 2002; 113:15-28.
51. Rogers LJ, Bolden SW. Light-dependent development and asymmetry of visual projections. Neurosci Lett 1991; 121:63-67.
52. Rogers LJ. Light input and the reversal of functional lateralization in the chicken brain. Behav Brain Res 1990; 38:211-221.
53. Manns M, Güntürkün O. Development of the retinotetcal system in the pigeon: A cytoarchitectonic and tracing study with choleratoxin. Anat Embryol 1997; 195:539-555.
54. Manns M, Güntürkün O. Monocular deprivation alters the direction of functional and morphological asymmetries in the pigeon's (Columba livia) visual system. Behav Neurosci 1999; 113:1257-1266.
55. Wong ROL, Gosh A. Activity-dependent regulation of dendritic growth and patterning. Nature Rev Neurosci 2002; 3:803-812.
56. Cohen-Cory S. The developing synapse: Construction and modulation of synaptic structures and circuits. Science 2002; 298:770-776.
57. Ruthazer ES, Cline HT. Insights into activity-dependent map formation from the retinotectal system: A middle-of-the-brain perspective. J Neurobiol 2004; 59:134-146.
58. Prior H, Diekamp B, Güntürkün O et al. Activity-dependent modulation of visual lateralization in pigeons. Neuro Report 2004; 15:1311-1314.
59. von Bartheld CS. Neurotrophins in the developing and regenerating visual system. Histol Histopathol 1998; 13:437-459.
60. Berardi N, Maffei L. From visual experience to visual function: Roles of neurotrophins. J Neurobiol 1999; 41:119-126.
61. Frost DO. BDNF/TrkB signaling in the developmental sculpting of visual connections. Prog Brain Res 2001; 134:35-49.
62. Vicario-Abejón C, Owens D, McKay R et al. Role of neurotrophins in central synapse formation and stabilization. Nature Rev Neurosci 2002; 3:965-974.
63. Pollock GS, Vernon E, Forbes ME et al. Effects of early visual experience and diurnal rhythms on BDNF mRNA and protein levels in the visual system, hippocampus, and cerebellum. J Neurosci 2001; 21:3923-3931.
64. Tropea D, Capsoni S, Tongiogi E et al. Mismatch between BDNF mRNA and protein expression in the developing visual cortex: The role of visual experience. Eur J Neurosci 2001; 13:709-721.
65. Gianfranceschi L, Siciliano R, Walls J et al. Visual cortex is rescued from the effects of dark rearing by overexpression of BDNF. Proc Natl Acad Sci 2003; 100:12486-12491.
66. Cohen-Cory S, Fraser SE. Effects of brain-derived neurotrophic factor on optic axon branching and remodelling in vivo. Nature 1995; 378:192-196.
67. Alsina B, Vu T, Cohen-Cory S. Visualizing synapse formation in arborizing optic axons in vivo: Dynamics and modulation by BDNF. Nat Neurosci 2001; 4:1093-1101.

68. Lom B, Cogen J, Sanchez AL et al. Local and target-derived brain-derived neurotrophic factor exert opposing effects on the dendritic arborization of retinal ganglion cells in vivo. J Neurosci 2002; 22:7639-7649.
69. Du JL, Poo MM. Rapid BDNF-induced retrograde synaptic modification in a developing retino-tectal system. Nature 2004; 429:878-883.
70. Theiss C, Güntürkün O. Distribution of BDNF, NT-3, rekB and trkC in the developing retino-tectal system of the pigeon (Columba livia). Anat Embryol 2001; 204:27-37.
71. Manns M, Güntürkün O. Differential effects of ocular BDNF-injections onto the development of tectal cells characterized by calcium-binding proteins in pigeons. Brain Res Bull 2005; in press.
72. Huang EJ, Reichardt LF. Neurotrophins: Roles in neuronal development and function. Annu Rev Neurosci 2001; 24:677-736.
73. Huang EJ, Reichardt LF. TRK receptors: Roles in neuronal signal transduction. Annu Rev Biochem 2003; 72:609-642.
74. Heumann R. Neurotrophin signalling. Curr Opinion Neurobiol 1994; 4:668-679.
75. Heumann R, Goemans C, Bartsch D et al. Transgenic activation of Ras in neurons promotes hypertrophy and protects from lesion-induced degeneration. J Cell Biol 2000; 151:1537-1548.
76. Arendt T, Gärtner U, Seeger G et al. Neuronal activation of Ras regulates synaptic connectivity. Eur J Neurosci 2004; 19:2953-2966.
77. Manns M, Güntürkün O, Heumann R et al. Photic inhibition of TrkB/Ras activity in the pigeon's tectum during development: Impact on brain asymmetry formation. Eur J Neurosi 2005; in press.
78. Hensch TK, Fagiolini M, Mataga N et al. Local GABA circuit control of experience-dependent plasticity in developing visual cortex. Science 1998; 282:1504-1508.
79. Robert F, Cuénod M. Electrophysiology of the intertectal commissures in the pigeon: Inhibitory interaction. Exp Brain Res 1969; 9:123-136.
80. Hardy O, Leresch N, Jassik-Gerschenfel D. Postsynaptic potentials in neurons of the pigeon's optic tectum in response to afferent stimulation from the retina and other visual structures. Brain Res 1984; 311:65-74.
81. Keysers C, Diekamp B, Güntürkün O. Evidence for physiological asymmetries in the physiological intertectal connections of the pigeon (Columba livia) and their potential role in brain lateralization. Brain Res 2000; 852:406-413.
82. Güntürkün O, Böhringer PG. Lateralization reversal after intertectal commissurotomy in the pigeon. Brain Res 1987; 408:1-5.
83. Güntürkün O, Hoferichter HH. Neglect after section of a left telencephalotectal tract in pigeons. Behav Brain Res 1985; 18:1-9.
84. Johnston AN, Rogers LJ, Dodd PR. [3H]MK-801 binding asymmetry in the IMHV region of dark-reared chicks is reversed by imprinting. Brain Res Bull 1995; 37:5-8.
85. Andrew RJ, Johnston AN, Robins A et al. Light experience and the development of behavioral lateralisation in chicks. II. Choice of familiar versus unfamiliar model social partner. Behav Brain Res 2004; 155:67-76.
86. Corballis MC, Morgan MJ. The biological basis of human laterality: Evidence for a maturational left-right gradient. Behav Brain Sci 1978; 2:261-269.

CHAPTER 3

Development of Vertebrate Brain Asymmetry under Normal and Space Flight Conditions

Alexandra Proshchina* and Sergey Saveliev

Abstract

We investigated the effects of spaceflight on the development of right-left brain asymmetry in larvae of amphibians (*Xenopus laevis*) and in pups and embryos of mammals (*Rattus norvegicus*). Here we report that larvae of *X. laevis* showed no changes in the volume of grey matter post exposure to microgravity, but did show increased volume of white matter and decreased volume of the retina, olfactory placodes and VIII cranial nerve ganglion size. Asymmetrical development of cerebral structures of *X. laevis* was, however, unaffected by spaceflight. Embryonic rats exposed to microgravity from days 9-20 of development had widespread neurodegeneration and, additionally, formed grey matter cavities. From a symmetry perspective, histological and morphometrical analysis found that space flight conditions reversed the typical left → right-sided enlargement of the nuclei habenulae and nuclei colliculi, and produced abnormalties in neural migration.

Introduction

Morphological asymmetries of the human brain, and their relationship to speech and thought, have been known for many years.[1] More recently, asymmetrical development of cerebral hemispheres in nonhuman, vertebrate animal species have also been well characterized,[2-4] although there has been debate in this area.[5] Here, we report on work that suggests that morphological asymmetry exists in the brain of common laboratory objects—the African smouth clawed frog, *Xenopus laevis*, and the rat, *Rattus norvegicus*, during normal development and, more importantly, that this development is influenced by the microgravity present during spaceflight.

Despite the fact that spaceflight has existed for almost five decades, little is known about the effect of microgravity on either the normal, or developing, brain. Spaceflight alters a host of normal peripheral physiological processes in the adult organism such as homeostatic control of the circulatory system, of mineral cycles and in tissue energetics.[6]

There are many reasons to believe that the developing organism may be more susceptible to the influences of microgravity than the adult. During development, ongoing processes such as active cell proliferation and formation of organ rudiments and systems are fragile. Instability of these processes under normal gravity conditions is well known, and numerous pathologies and abnormalities of development are found in nature. At the same time, mineral cycle and tissue metabolism in the developing animal are very flexible compared to in

*Corresponding Author: Alexandra Proshchina—Institute of Human Morphology RAMS, ul. Tsurupi, 3, Moscow 117418, Russia. Email: proschina@mtu-net.ru

Behavioral and Morphological Asymmetries in Vertebrates, edited by Yegor B. Malashichev and A. Wallace Deckel. ©2006 Landes Bioscience.

adults. Whether or not these differences lead to differential susceptibility to microgravity is unknown, but possible. For this reason, we investigated morphological right-left brain asymmetry in larvae of the amphibians *X. laevis* and in embryos of the mammals *R. norvegicus* in normal and spaceflight conditions.

The Effects of Spaceflight on Brain Development in Frog

This investigation was a part of the Russian - Canadian "Development" experiment, which flew in space on the BION-10 satellite. It was devoted to studying the influences of microgravity on the brain and sensory system of *X. laevis* larvae.

Materials and Methods

Adult females of *X. laevis* bred at the Department of Embryology of Moscow State University (MSU) were mated with males from Nasco (Fort Atkinson, Wiskonsin) commercial shipments. Ovulation, amplexus and fertilization were initialized with human chorionic gonadotropin according to the standard method generating a few hundred fertilized eggs. Of these, 106 eggs were divided into two groups.

The first group of 53 eggs was marked as 'space animals'. They were placed into a three-liter rectangular plexiglass containers, commonly used on the Biokosmos satellite, half filled with boiled aerated standing tap water at 21.5°C. After obtaining measurements of oxygen, hardness and the pH of water, stage[7] 6.5-7 larvae were added to the water and the container was sealed by a metal plate with a rubber hermetic. Forty hours later, when the tadpoles were presumably at stage 25 (according to observations of control group, see below) the eggs were launched. The container remained sealed until the end of experiment 11.5 days later.

The control group, also consisting of 53 animals, was used as a 'synchronous control'. The embryos were prepared and contained in a capsule similar to the launched one. The temperature on the satellite was maintained constant during the flight and the synchronous control group was kept at the same temperature. Although the organisms were not fed during the experiment, it is possible that they consumed microorganisms which developed in the surrounding water.

The flight container and the container of synchronous controls were opened at room temperature two hours after landing. All of the animals were at stage 47 of development, which corresponds to young free Y N"ming feeding tadpoles. Organisms from each group were fixed in 4% neutral buffered formaldehyde. Of these eight launched and 10 synchronous controls were studied histologically in the Institute of Human Morphology. Specifically, the material was dehydrated in a series of alcohol solutions of increasing concentration, dioxaned and embedded in paraffin. The paraffin blocks were serial sectioned at 10 μm and the sections were stained with Mallory's method.

Graphical reconstructions of the head were prepared with the aid of microprojector. In *X. laevis* larvae we studied the volume and surface area of the brain (grey matter and white matter) and peripheral analysers (olfactory placodes, retina, and ganglion of the VIII-th nerve) on the right and on the left. The volume of the brain and its parts was counted according to the formula (1).[8] The area of the sections was measured on the drawings made with the projection apparatus at a certain magnification.

$$V = \sum \frac{Sn \times m \times d}{D^2} \tag{1}$$

Where: V = the volume of brain or its part
Sn = area of the measured section
m = the number of section in the sequence
D = linear magnification
d = the thickness of the sections

The surface of the brain and its parts was measured in a complete series of sections calculated according to the formula (2):[8]

$$Z = \sum \frac{P \times m \times d}{D} \tag{2}$$

Where: Z = area of the surface
P = perimeter of the measured section
m = the number of section in the sequence
D = linear magnification
d = the thickness of the sections

The area of each structure was measured with the aid of the planimeter, the perimeter—with the aid of the curvimeter.

We calculated asymmetry in the brain according to the formula (3):[9]

$$A = \frac{R - L}{R + L} \times 100 \tag{3}$$

Where: A = asymmetry of a structure
R = the size of the structure on the right side
L = the size of the structure on the left side

Statistical analysis was carried out with the aid of the program Statistica for Windows 4.3 (Stat Soft Inc.) using a non parametric analysis (Wilcoxon's test).

Results and Discussion

The volume and surface of the whole brain in control and experimental groups were similar. The volume of grey substance was also the same in both groups. The volume of white substance in the microgravity-exposed space group was 30% greater than in controls. Conversely, the launched subjects showed reduced volume in the retina, olfactory placodes and VIII nerve ganglions (60%, 21%, and 22% respectively). The right-side asymmetry of investigated structures was preserved in the launched group and amounted to 2-9% (Fig. 1). Thus, while symmetrical development of the brain was preserved in the microgravity condition, spaceflight nonetheless altered whole brain development.

In amphibians, morphological brain asymmetry first appears at neurulation stage.[10] This asymmetry was retained during all post-embryonic (larval) stages, both in launched and in control groups, in the retina, the ganglion of the 8th nerve and the olfactory placodes. The cause of the changes in the grey/white matter ratio remains unclear, but is likely due to effects of space flight on morphogenesis of the nervous system.

The Effects of Spaceflight on Brain Development in Rat

This investigation was a part of the NIH.R1 mission jointly sponsored by the NASA Life Sciences Division and the National Institutes of Health. The "Rodents 1" payload included 11 experiments conducted by scientists from the USA, France, and Russia. The objective of the experiments was to investigate the role of gravity in developmental processes, particularly, in establishment of right-left asymmetry. The latter work was carried out by us together with the Institute of Medical and Biological Research (Moscow, Russia).

Materials and Methods

Four groups of *R. norvegicus* (Sprague-Dawley strain; Taconic Farms, Germantown, NY) were studied, including: (1) newly born rats, which developed during spaceflight and delivered after landing (flight group, F), (2) newly born rats, which developed on Earth under conditions, except for the gravity, similar to the space flight group (synchronous control group, SC), (3) newly born rats, and (4) E20 embryos, which developed in standard vivarium conditions (vivarium control groups, VC-NB and VC-E correspondingly).

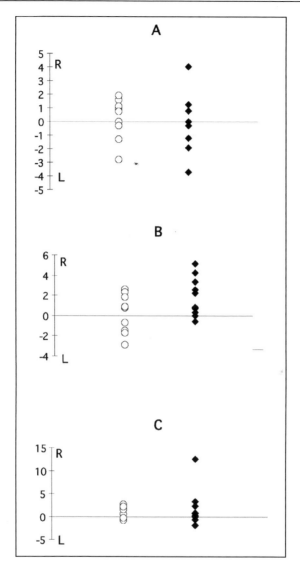

Figure 1. The asymmetry of the brain in *X. laevis* larvae. Coefficients of asymmetry (see Materials and Methods, formula 3) in volumes of grey (A) and white (B) matter, and brain surface area (C). R: right-, L: left-sided asymmetry. Empty circles: controls; filled rhombuses: space flight group.

For the spaceflight condition, 10 pregnant dams (F group) onboard the US Space Shuttle were subjected to space flight from days E9 to E20. These times corresponded with developmental stages during which neurulation and organogenesis take place. As rats typically give birth on day 22 of gestation, the developing rats were subjected to microgravity for the majority of their embryonic and fetal life.

Specifically, rats were initially shipped to Kennedy Space Center on the second day of gestation and placed in standard vivarium cages. The dams were subjected to laparotomy on the seventh day of gestation to establish the number of implantations. Ten dams, each with at least five implantation sites per uterine horn, were selected for flight. These rats were loaded into two Animal Enclosure Modules (AEMs) on the next day, and the AEMs were then placed in the middeck of the Shuttle (see NASA web-site for more details on AEMs: http://lifesci.arc.nasa.gov/lis2/Chapter4_Programs/NIH_R/NIH_R1.html). All pups were born two days after their return to Earth. One pup was taken by us from each dam.

The delayed synchronous controls (SC group) were treated in a manner identical to the flight animals, following a 24-hour delay. The group was housed in AEMs within the Orbiter Environmental Simulator (OES). The OES is a modified environmental chamber at Kennedy Space Center whose temperature, humidity, and CO_2 level are electronically controlled based on downlinked environmental data from the orbiter. Thus the animals within the chamber are exposed to environmental conditions that are similar (except for gravity) to those experienced by the flight group during the mission. These controls were subjected to laparotomy at the same time as spaceflight group. Again, one pup from each female was studied histologically.

The rats in the vivarium control groups were individually housed in standard vivarium cages. Vivarium control animals did not undergo laparotomies, as the rats in the flight and delayed synchronous control groups did. Seven NBC were delivered by natural birth. Seven embryos at E20 were obtained by dissection of the uterus horns.

Rat brains were processed by methods similar to those described for the *Xenopus* larvae. After decapitation of isosulfanized animals, brain specimens were fixed in Bouin's fixative, embedded in paraffin, serially sectioned at 15 μm thickness. Every fifth section was stained with haematoxylin-eosin or with the Mallory method.

In rat brains we focused attention on five structures: superior (SQC) and inferior quadrigeminal colliculi (IQC), medial (MHN) and lateral habenular (LHN) nuclei and caudate nucleus (CN). These structures were chosen because they belong to different brain divisions (IQC and SQC—to mesencephalon, LHN and MHN—to diencephalon, CN—to telencephalon). This allowed estimating the experimental influence of the space flight on various brain divisions. The nuclei are also included in different functional complexes of the brain: SQC are a part of the visual system, IQC—auditory system, while LHN and MHN display extensive liaisons with the limbic system and interact with olfaction and epiphysis. The CN features functional links with the limbic system and is the target of the nigrostriatal system. Apart from this the structures discussed are markedly recognizable and are large, making them suitable for the stereotaxic analysis. The margins and location of nuclei were identified according to the atlas of rat brain.[11]

To study the asymmetry in the rat brain the equipment of the Wacom Computer Systems GmbH-Wacom Ultra Pad was used (Ultra Pen Duo Stylus and Ultra Pad A5). The areas and perimeters of structures were measured with the computer program Canvas 5.02 (Deneba Systems Inc.).

The volume and the surface of investigated structures, and the coefficient of asymmetry were calculated according to the formulas 1-3 (see above). Statistical analysis included a non parametric analysis (Wilcoxon's test) carried out with the aid of Statistica for Windows 4.3 (Stat Soft Inc.).

Results and Discussion

Evidence of Neurodegeneration in Microgravity Pups

Histological analysis of pups from the flight group revealed neuronal degeneration in 70% of the subjects, with cell loss found in all of the brain regions studied (e.g., telencephalon, see Fig. 2). Although SC subjects showed some defects in the brain (Table 1), no degenerative changes were found there. Additionally, the brains of the spaceflight group formed cavities of degeneration and showed separation of the white matter. The areas of neurodegeneration were about 50 to 400 μm in size. No evidence of any neuroglial proliferation or scars in the form of neuroglial or collagenous fibres were seen in these regions of degeneration. The defective areas of tissue were often seen next to the zones of normally

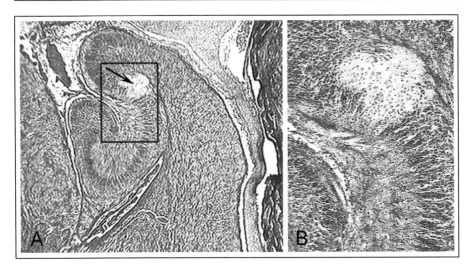

Figure 2. Serial frontal sections through the telencephalon of *R. norvegicus* P0 pups (flight group, F). Mallory staining. A centre of neuron degeneration (arrow) is shown at low (A) and a higher (B) magnification; the rectangular on (A) highlights the region shown on (B).

Table 1. Results of the microscopic study of the brain in rat pups in conditions of spaceflight and in Synchronous control group

	Flight Group	Synchronous Control
1. Pups with no defects of the brain	3	8
2. Pups with defects of the brain	7	2
3. Pups with defects of the telencephalon	5	1
4. Pups with defects of the diencephalon	3	1
5. Pups with defects of the mesencephalon	4	0
6. Pups with defects of the metencephalon and myelencephalon	3	0

developed neural tissue. The lack of neuroglial proliferation in these areas of neurodegeneration suggests that the pathology may have developed in very early embryos.

Changes in Auditory, Visual, Habenular and Caudate Brain Regions

Auditory Systems

In mammals the fibers of the 8-th nerve are projected into the cortex through the medial geniculate body and branch collaterals into the inferior quadrugeminal colliculi (IQC). Since inferior colliculi not only receive collaterals but themselves give off fibres into the medial geniculate body, we can presume that asymmetry of the posterior colliculi can be related to asymmetry of the auditory and vocal zones of the cortex.

The IQC are significantly larger on the left side in normal pups (SC and VC-NB subjects), whereas in pups developed under spaceflight conditions (F) and in normal E20 embryos (VC-E) the right-side IQC was larger in size (Fig. 3A).

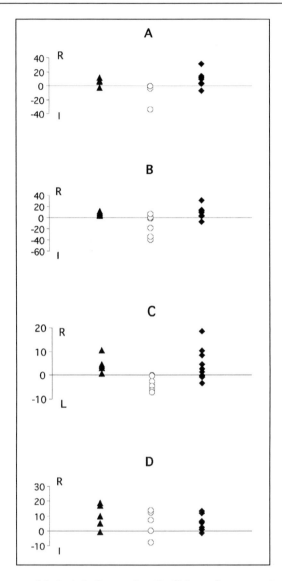

Figure 3. The asymmetry of the brain in *R. norvegicus*. Coefficients of asymmetry in volumes of inferior (A) and superior (B) quadrigeminal colliculi, and in surface area of the medial (C) and lateral habenular (D) nuclei. R: right-, L: left-sided asymmetry. Filled triangles: vivarium controls (VC-E, E20 embryos, N = 7); empty circles: synchronous controls (SC, P0 pups, N = 10); filled rhombuses: space flight group (F, P0 pups, N = 10).

The spaceflight-exposed rats did not, as a group, show changes in their auditory system.[12] However, we found inverted asymmetry of the IQC in F group subjects. It is unclear if these changes are caused by a direct influence of microgravity, or from nonspecific effects related to the conditions of spaceflight. Indeed, the pattern of asymmetry in the pups subjected to the flight conditions is more similar to that in normal E20 embryos, than to control P0 pups. Moreover, Serova,[6] who also participated in these experiments, found a decrease in tactile and vestibular sensitivity in the newborn spaceflight-exposed subjects. In P5 pups, however, all difference between the experimental and control animals disappeared, suggesting postexperimental recover.

Visual Systems

The superior quadrigeminal colliculi (SQC) are part of the visual system. In mammals the visual images are projected through the lateral geniculate body into the cortex. The fibers of the optical nerve end inside the lateral geniculate body, and their collaterals are directed into the optical layer of the anterior colliculi of the quadrigeminae. In many mammals the SQC develops asymmetrically with dominance of the left anterior colliculi. Our results show that in VC-NB pups the left anterior colliculi are also larger than the right ones. However, in the F group, subjected to microgravity, the asymmetry of the anterior colliculi is reversed. The volume of the colliculi on the right side was reliably larger than on the left side (Fig. 3B). Note, however, again the right-side pattern of the SQC size in E20 control embryos (see also discussion above).

The retina of experimental animals in the spaceflight group was thinner, probably as a consequence of a thinning out of bipolar retinal cells.[12] The functional properties of such changes in retinal structure probably stay in the adaptive limits of the "reaction norm." However, whether such changes can significantly affect vision during long flights in fetal or adult subjects is not known.

Habenular Nuclei

The structural asymmetries of the Diencephalon has long being known since the work by Breightenberg and Kemali,[13] and Morgan.[14] The habenular are connected in development with the parietal eye. In mammalian species with reduced pineal organ the habenular nuclei are involved into the olfactory system, having symmetrical anatomy.

In the control pups (SC, VC-NB) the left MHN was larger. Conversely, in E20 vivarium control embryos and F pups it was the right side that was larger (Fig. 3C). The general volume of the medial habenular nuclei was 21% less in F pups, which developed while their mothers were subjected to microgravity. A plausible explanation for this, much like in the case of auditory and visual systems, is an underdeveloped status of nucleus in the flight group. In LHN a significant rightward asymmetry was found in F and VC-E groups. Here, however, SC pups also demonstrated the same direction of asymmetry, although this did not reach significance (Fig. 3D). This may indicate a lesser effect of the space flight on the development of the LHN.

Caudate Nucleus

It was noted above that the caudate nucleus fulfils a number of integrative functions and constitutes a part of the nigro-striatal and extropyramidal systems. In man, the asymmetry of the caudate nucleus is well expressed and larger on the right.[15,16] No asymmetry of the caudate nucleus, however, was found in any of our subjects.

Conclusions

Histological and morphometrical analysis of the brain and peripheral analyzers revealed retardation of the brain development and, in some cases, reversion of the brain asymmetry during space flight. These changes are seen particularly in major sensory systems, including the visual, acoustic and olfactory systems. The left-side asymmetry in brain structure of the control rat pups (nuclei habenulae laterales et mediales, nuclei colliculi inferiores et superiores) changed to the right-side asymmetry in microgravity. Interestingly, the embryonal stages, as could be seen from vivarium controls, showed the right-sided asymmetry in the studied brain structures. This raises a possibility that the observed asymmetry reversion is due to greater underdevelopment of the brain in rat pups developed in space. The results from the experiment with amphibian larvae confirm the overall retardation of the brain development under conditions of microgravity. In *Xenopus*, however, the normal pattern of asymmetry was retained, indicating a greater sensitivity for the brain in rats in comparison with the amphibians. Our experiments raise a concern about the safety of long term spaceflight in impregnated mammals. The findings clearly indicate that microgravity, and possibly other nonspecific effects of spaceflight, can alter normal development of the brain. Whether or not these findings will occur in humans remains an untested, but important, consideration in the planning of future long-term space missions in which both men and women are present.

References

1. Geschwind N, Levitsky W. Human brain: Left-right asymmetries in temporal speech region. Science 1968; 161:186-187.
2. Bianki VL. Asymmetry of the brain of animals. Leningrad: Nauka 1985.
3. Springer S, Deich G. Left brain, right brain. Moscow: Mir, 1983.
4. Filley CM. Neurobehavioral anatomy. Niwot: University Press of Colorado, 1995.
5. Deglin VL. Lectures on functional human brain asymmetry. Amsterdam, Kiev, 1996.
6. Serova LV. Adaptive ability of vertebrates in condition of weightlessness. Aviacosm and Ecol Med 1996; 30(2):5-11 (In Russian).
7. Nieuwkoop PD, Faber J. Normal table of Xenopus laevis. Amsterdam: North Holland, 1956.
8. Blinkov SM, Glaser II. Brain in figures and tables. Moscow: Medicina, 1964.
9. Bullmor E, Ron M, Harvey I et al. Agaunst the laterality index as a measure of cerebral asymmetry. Psychiatry Res 1995; 61(2):121-124.
10. Proshchina AE, Saveliev SV. Study of amphibian brain asymmetry during normal embryonic and larval development. Izv Akad Nauk Ser Biol 1998; 25(3):408-411.
11. Pellegrino LJ, Pellegrino AS, Cushman AJ. A stereotaxis atlas of the rat brain. New York: Appleton-Century-Crofts, 1979.
12. Saveliev SV, Serova LV, Besova NV et al. The influence of weightlessness on the neuro-endocrine systems development. Aviacosm and Ecol Med 1998; 32(2):31-36.
13. Braitenberg V, Kemali M. Exceptions to bilateral symmetry in the epithalamus of lower vertebrata. J Comp Neurol 1970; 138:137-146.
14. Morgan MI, O Donnel, Oliver RF. Development of left-right asymmetry in the habenular nucleu of Rana temporaria. J Comp Neurol 1973; 149:203-214.
15. Castellanos FX, Rapoport JL, Hamburger SD et al. Quantitative morphology of the caudate nucleus in attention deficit hyperactivity disorder. Am J Psychiatry 1994; 151(12):1791-1796.
16. Raz N, Acker JD, Torres IJ. Age, gender, and hemyspheric differences in human striatum: A quantitative review and new data from in vivo MRI morphometry. Neurobiol Learn Mem 1995; 63(2):133-142.

Is There a Link between Visceral and Neurobehavioral Asymmetries in Development and Evolution?

Yegor B. Malashichev*

Abstract

Behavioral laterality on the basis of physiological neural asymmetries does not seem to develop under the control of the same developmental mechanisms as asymmetries of the visceral organs. Earlier, we have found little evidence linking these two groups of asymmetries, which implies different developmental regulatory pathways and independent evolutionary histories for visceral and telencephalic lateralizations.[1] In this Chapter further arguments are considered supporting independent developmental and evolutionary pathways for visceral and neurobehavioral asymmetries in vertebrates. Although the question remains contradictory in view of some new evidence, this review implies, in particular, that the search for developmental mechanisms and genes controlling the establishment of brain lateralization (e.g., differential functioning of telencephalic hemispheres) can be based on approaches, which differ from attention only to human subjects and/or pathways leading to asymmetric morphologies in the diencephalon and major viscera. Recent advances in the studies of asymmetries in invertebrates reveal deep roots for both neurobehavioral and visceral asymmetries dating the history of directional asymmetries back to the earliest bilateral organisms. This makes the understanding of developmental interactions between different asymmetry types complicated, but on the other hand, also makes possible diversification of the experimental subjects and experimental approaches.

Introduction

Asymmetries of the vertebrate body (e.g., visceral organs) and the head (brain) have a number of features, which may reflect their different evolutionary history and, probably, different developmental pathways. Recent reviews of developmental aspects of vertebrate asymmetries pointed to these intriguing phenomena.[1-3] In particular, Malashichev and Wassersug[1] showed that virtually no evidence had been reported to date on any strong linkage between the development of visceral asymmetries and that of lateralized functions of the brain hemispheres in any vertebrate class (similar discussion in ref. 2). We further discussed those features of the visceral and behavioral asymmetries that assign them to two mostly independent categories. For instance, a point of divergence is in that behavioral asymmetries can be lateralized or not, showing a greater degree of variation, whereas the visceral asymmetries are usually lateralized, demonstrating one-sided population alignment common for all vertebrates (see also ref. 3).

*Yegor B. Malashichev—Department of Vertebrate Zoology, St. Petersburg State University, Universitetskaya nab., 7/9, St. Petersburg, 199034, Russia. Email: malashichev@gmail.com

Behavioral and Morphological Asymmetries in Vertebrates, edited by Yegor B. Malashichev and A. Wallace Deckel. ©2006 Landes Bioscience.

Furthermore, the genetic background of behavioral lateralization is less pronounced and less understood than that of visceral asymmetries. Here I analyze in further details the issues raised in the previous reports and show the perspectives, which as I believe, the studies of development and evolution of the brain asymmetries may have.

To make deeper insight into the evolution and development of morphological and physiological brain asymmetries a number of approaches may exist. One way is to investigate in a comparative manner behavioral lateralization in diverse groups of organisms, which could develop asymmetries either in a common or much a different way one from the other. Another approach, which seems now very promising, is to look for new gene cascades controlling the development of most neurobehavioral asymmetries, which might be different from the well-known Nodal-cascade leading to asymmetry of the visceral organs and diencephalon. Combining and using both of these approaches might be considered as the best way for addressing the developmental and evolutionary origin of neurobehavioral lateralization.

Upside Down: Hydra Model

Meinhardt[4] suggested a simple model of head and trunk evolution in two distinct animal lineages, namely vertebrates and insects. Based on the comparison of spatial expression of regulatory genes responsible for early embryo patterning (see ref. 4 for more details), this model claims that the whole body of a vertebrate or an insect is homologous only to the aboreal part (oral opening and the disk with tentacles) of a cnidarian (e.g., hydra). The head, instead, in both animal classes is homologous to the rest of the hydra's body. Although having a bit speculative character, this model is testable and assumes important conclusions.

First, the body in vertebrates and insects evolved differently (e.g., the development of the embryonic midline).[4] It is suggestive therefore, that the establishment of the embryo's left-right polarity in these two highly divergent groups may involve different developmental mechanisms (see an extensive discussion from variety of grounds in refs. 1,3). It is noteworthy to add here that homologues of *Drosophila* genes regulating left-right axis polarity, heart and gut looping are not localized to the 6p21, 6q14-q21, 6q25, 7q22, 9q32-q34, 10q21-22, 11q13, 11q25, 12q13, 13qter, or Xq24-q27.1 chromosome regions highlighted by heterotaxic patients or mutant mice with visceral asymmetry defects (see OMIM database and refs. 5-7).

Second, the head and the trunk could acquire their own developmental mechanisms from the earliest evolutionary stages.[1] Indeed, since early works on handedness in humans with left-right visceral reversions (situs inversus) there have being no reports of strong correlates between the visceral situs and asymmetric neurobehavioral traits with exception to a few brain morphologies.[8,9] This lack of clear correspondence between the two kinds of asymmetries in humans was further supported by observations of different patterns of head and trunk asymmetries in conjoined and nonconjoint twins[11] (all facts together summarized in refs. 1-3, and 12).

Cooke[3] suggested that visceral asymmetries found in vertebrates may be descendants of asymmetries in the gastric cavity of some echinoderms. He further speculated that the vertebrate origin is best understood in terms of novel imposition of otherwise bilateral neural features and their further cooptation to the ancestral asymmetric "visceral" animal. Although this model by Cooke is supportive and implies that the vertebrate visceral situs might be a pedigree of the asymmetric visceral body of echinoderms, we[1] argued that both visceral and neurobehavioral asymmetries could evolve in parallel from the earliest steps of chordate evolution and most probably predate the origin of Chordata. Furthermore, it is possible now to put the origin of both neurobehavioral and visceral asymmetries even earlier and outside not only chordates, but even deuterostomes, suggesting that the origin of both types of asymmetry coincide with the emergence of the bilateral symmetry itself; the proposition, which is difficult but, nevertheless, possible to prove.

Perhaps, the reality is that neurobehavioral asymmetries emerged early in evolution, but were later indeed superimposed on the visceral asymmetries in Chordata, Echinodermata and close ancestors; hence providing multiple examples of correlation (due to functional integrity of a complex organism), but nonetheless making not self evident a developmental program common for both kinds of asymmetries.

Our rough notion on the lack of asymmetric body morphologies in majority of bilateral invertebrates (with exception of a few classical examples like late developing claw asymmetry in crustaceans, and keeping completely aside the spiraled shells of mollusks and brachiopods; see ref. 13 for a review) has been broken by the report that the flat worms, which otherwise possess symmetric bilateral body are asymmetric in their eye-regeneration response in the presence of a H$^+$/K$^+$-ATPase inhibitor.[14] The H$^+$/K$^+$-ATPase activity has already been shown to be important for establishing asymmetry in the electric cell membrane potentials in the embryo and, as a consequence, the visceral asymmetry formation in vertebrate species,[15] some tunicates (nonvertebrate chordates),[16] and echinoderms.[17] Therefore, these new information dates back the history of morphological/physiological asymmetry to planarian (see Fig. 1 for phylogenetic position), one of the earliest bilaterian organisms. It also supports probably a shared ancestral mechanism of morphological asymmetry formation for all Bilateria.

Interestingly, there are other facts that give a basis for thinking of the vertebrate situs as a phenomenon, which has long being preexisted in a somewhat "hidden" state. For example, in *Hydra* the transfer of fluid by the peduncle has a similar neurological and genetic basis to the pumping of blood by the heart in vertebrates.[18] Although no asymmetry was shown in the *Hydra* peduncle, the possibility that it shares with the vertebrate heart a common ancestral origin, in couple to the existence of physiological left-right asymmetry in planarians suggests also an early evolutionary opportunity for the emergence of asymmetry comparable to that in the vertebrate heart in the homologous pumping organs of other bilateral organisms. We, however, have no better knowledge of this intriguing question and are entirely in the dark about the concrete mechanisms, by means of which the physiological asymmetry like that described in

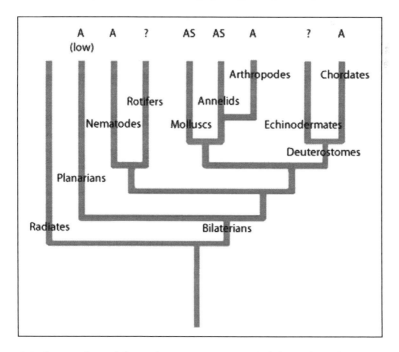

Figure 1. Distribution of neurobehavioral asymmetries in animal phyla. A: directed (population level) asymmetry; AS: antisymmetry; S: symmetry; ?: no information. The phylogenetic tree is based on schemes from Raven PH et al. Biology, 7th Edition. MaGraw-Hill, 2005.

flat worms became manifested in a more visualized structural forms. Equally, we do not know whether the morphological left-right asymmetry evolved gradually in bilateral animals or has long being existed as a difference in physiological reactivity between the two sides of the body, just providing an operative basis for rapid independent evolution of morphological asymmetry in certain animal phyla, e.g., echinoderms and vertebrates.

Thus, studying different (especially phylogenetically ancient) bilaterian and even radiate animals will probably give us more information on the early evolution of morphological left-right asymmetry. To answer the question on whether morphological (heart and visceral) asymmetries are invariantly coupled to neurobehavioral lateralization, it would be also important to learn, whether the latter possesses comparably long evolutionary history. One interesting model could be arthropods, and particularly, insects. I have already mentioned that the midline formation in the body of vertebrates and insects differ drastically between these two groups (see also refs. 1,4). This difference probably leads to the typical left-right body asymmetry in vertebrates and the spiral body formation in insects, without clear differences between the sides.[1] However, the head (or more precisely, the brain) and neurobehavioral asymmetries are not necessary completely different in two groups under discussion, but rather could principally form a parallel similar pattern in these proto- and deuterostome phyla.[1] Therefore, a possibility exists, that neurobehavioral asymmetries, normally associated with vertebrates may be found in arthropods, and even in other invertebrate groups as early as first bilaterians.

In the next section I summarize and discuss some scarce and rarely cited, and, therefore, less known data on the distribution of neurobehavioral asymmetries among invertebrates. These data provide arguments in support of asymmetry early evolution on one hand and, nevertheless, of their independence from the side differences, which lead to the morphological asymmetries of the animal body (e.g., in the heart and the gut) on the other hand.

Nonvertebrate Neurobehavioral Lateralizations

Whether invertebrate animals possess any neurobehavioral asymmetries? Given that insects have dual brain, which is at least partially split into left and the right halves and have paired contralateral structures (e.g., optic lobes), it is reasonable to expect that principally these halves can have different specializations. Social insects are of our immediate interest. Recently, Vallortigara and Rogers[19] have argued that the alignment of the direction of behavioral asymmetries at the population level takes place under "social" pressures occurring when individuals must coordinate their asymmetric behaviors within the species or when interacting with other species (e.g., prey or predators). If this generalization is true, then we should expect a greater level of lateralization in social species (and it is indeed so in fish),[20] and therefore it might be an easier task to detect such a lateralization in a social, rather than a solitary insects to be a primary estimate on whether invertebrates are lateralized or not.

The classical object of behavioral research since early K. von Frisch work[21,22] are honeybees (*Apis mellifera*). Whether the honeybees possess anatomical or functional brain asymmetries and/or behavioral lateralizations is intriguing,[1,3] but unfortunately, mostly unexplored issue. What we definitely know is that bees are able to discriminate left from right during olfactory learning[23] and are capable of perception the symmetric vs. asymmetric structures.[24,25] To the best of my knowledge, there are no studies, which would show any population-level lateral bias in turning, flight or coordination with other individuals in honeybees. However, this may indicate not a lack of corresponding behavior, but rather may reflect the scarcity of attempts aimed to reveal any neurobehavioral asymmetry in honeybees, because other species of social Hymenoptera, e.g., bumblebees and ants, do possess lateralized behaviors.

For example, in bumblebees of four species (*Bombus lapidarius*, *B. terrestris*, *B. lucorum*, and *B. pascuorum*) the direction in which they rotated around inflorescences of *Onobrychis viciifolia* (Fabaceae) was scored.[26] Three species (except for *B. terrestris*) have shown in this study a stable and significant preference to rotate either clockwise (*B. lapidarius*) or anticlockwise (the rest two species) when visiting flowers. However, based only on color pattern field identification of *B. terrestris* and *B. lucorum*, which demonstrated generally opposite direction of rotation, is difficult and not fully reliable. It was not possible, therefore, to exclude that nonsignificant result for *B. terrestris* was due to the mixture of two species in one sample and not due to the lack of lateralization in the flower visits; the suggestion that, however, remains to be proved.

As early as in mid 80s of the last century, in a series of pioneering reports on learning in ants (*Mirmica rubra*) in a multiple-dichotomy maze, Udalova and Karas[27,28] showed a substantial individual and population-level spatial motor asymmetry in this species in conditions of social motivation (alimentary or descendants care). In conditions of food motivation the total number of movements in the right side of the maze was significantly greater than in the left one, with 70% of individuals in the ant population showing significant preferences. The same significant rightward tendency was observed in approaches to the goal-point with reinforcement and first turns in the labirynth.[28] In labyrinth search trials, in which the ants approached the reinforcement goal-point the lateralization increased. The similar results were obtained in conditions of other type of social motivation—parental care (transportation of larvae).[27] These observations allowed authors judging on existence of species-specific (i.e., population-level) lateralization of motor-spatial behavior in ants, which was higher in conditions of social motivation.

Ants of another species, *Lasius niger*, were also shown to keep more to the right on tree trails, to expose the left side of the body to the closest nest-mate, and to turn mainly to the right in usual, while to the left—in "alarm" conditions.[29] Although the latter report contains data, which are more preliminary than desired, altogether, the cited evidence indicate some interesting analogies in the lateralization pattern in ants and higher vertebrates. Indeed, the rightward biases to orient themselves in the maze or in the open field experiments were found in mice and rats.[30,31] Left sided inspection of the conspecifics, and the differences in lateralization during slow and fast reactions to the external stimuli are also described in variety of vertebrates from fish to mammals (see for instance refs. 32-35 and Chapter 8 by A. Robins in this volume). The neural basis for spatial and cognitive asymmetry in ants, is not yet known and may be likely based on existent taxes and asymmetrical distribution of receptors in antennas. However, one common rather than two unique mechanisms of brain functioning in vertebrates and ants should not be overlooked.

Few other insect species were put on trials in searches for brain asymmetry or its consequences—lateralization of motor or sensory response. At least functional neural asymmetry was definitely shown in cockroaches (*Nauphoeta cinerea*) and fruit flies (*Drosophila melanogaster*). Females of the former species reduced the ability to find a male and mate after amputation of the right, but not the left antenna (right-side olfactory blindness).[36] This was due to the organization of the deutocerebrum, where a pronounced population level asymmetry in distribution of the glomeruli was found with more glomeruli on the right side of the olfactory lobes. Recently, anatomical brain asymmetry in fruit flies was shown to be associated with normal memory processing.[37] Here, a previously unknown structure was found at the right side of the brain of majority of flies in the population, while the minority group revealed its symmetrical position in both brain hemispheres, and as a consequence, the lack of long-term memory. Interesting, that hemispheric asymmetry in prefrontal cortex (PFC) and hippocampus activities is also an important brain characteristic involved in long-term memory processes in humans. The ability to learn and remember new information declines with aging and PFC activations in elders tend to be less asymmetric, which is more likely a compensatory mechanism involving the opposite hemisphere, rather than the primary source of long-term memory loss.[38] However, the very principle of using brain hemispheres asymmetrically for remembering and retrieval of information in humans and flies seems to be similar.

Faure and Hoy[39] further presented a nice and interesting synopsis of known facts on the asymmetry in the insect auditory system, although expressing some doubts on the relevance of methods used in measuring auditory lateralization. Faisal and Matheson studied righting behavior of desert locusts (*Schistocerca gregarina*).[40] These insects were overturned on its back in conditions of induced thanatosis (quiescence), which allowed orienting the legs of the insect in a controlled and symmetrical start position. As locusts like frogs have very long hind legs, which can touch the substrate when the animal is in upside down position, the authors interested on whether locusts can turn themselves into the normal position and have a lateral bias like that the frogs have (see ref. 41 for more details on amphibian lateralized righting). Indeed, the locusts righted themselves very effectively and independently, also demonstrating individual preferences to right themselves to either side. Lateralization at the population level was, however, not found. On my opinion, this lack of population level lateralization in hindlimbs actually agrees with the symmetrical mode of locomotion utilized by locusts. Indeed, frog species, which prefer alternating-limb locomotion, do show population level lateralization in righting themselves on a horizontal surface. However, those species, which use both hind legs simultaneously for effective jumping, much like jumping locusts, consequently possess lower population level motor lateralization than those using alternating-limb walking mechanics.[42]

The above conclusion can be confirmed with examples came from spiders.[29,43] Here the preferential use of the left foreleg was shown in a spider *Scytodes globula* in situations of touching and handling the prey.[43] The same tendency was noted in a few *Tegenaria atrica* spiders.[29] The number of left-leg lesions here was also significantly greater than the right-leg lesions. As touching and handling prey imply specialization of legs and their asymmetrical use, these findings are very important to understand that the functional asymmetry of the neural system is not a unique characteristic of vertebrates and that it might be greater or less dependent on the necessity of asymmetrical action.

More examples of population level motor asymmetry in invertebrates are now available. For example, spatial motor lateralization for the preference to swim in the clockwise direction was found in shrimps (*Gammarus oceanicus*) in the open-field test.[44] In mollusks (*Octopus vulgaris*), strong individual preferences for monocular viewing of the food model were found very recently.[45] In this species 23 of 25 tested individuals demonstrated significant side preferences, although the overall distribution of left versus right-eye subjects in the sample was anti-symmetrical. Another study revealed a pronounced asymmetrical effect of γ-aminobutyric acid (GABA) injection on postural orientation in a marine mollusk, *Clione limacina*, which leads to more leftward swims due to greater inhibition of the right motor neurons and interneurons.[46] This effect, found in mollusks is striking, given the role of GABA and GABA-receptors in processes of normal light induced behavioral asymmetry formation in birds (see, for instance, Chapter 2 by M. Manns in this volume) and pathological conditions, such as schizophrenia in man (e.g., see ref. 47).

There are also examples of neural (e.g., sensory or motor) asymmetries even in three completely different types of worms. Nematodes, *Caenorhabditis elegans*, are becoming a model of asymmetry development.[11,48,49] These worms possess several types of asymmetrical structural and functional features in neuronal organization, including left-right asymmetric gene expression patterns. Left-right antisymmetry of neuronal development in consequent body segments was described in leeches (Annelids).[50] Finally, of five species (totally 1768 individuals) of flatworms (Planarians),[51] which were tested for lateralization of turning in T-maze, three species demonstrated right-sided and one species—left-sided population-level turning preference. In contrast to vertebrates the number of lateralized individuals in flatworm samples varied from one third to one half, thus indicating lower overall level of behavioral lateralization, perhaps, reflecting an early step of evolution of neurobehavioral asymmetries.

Figure 1 summarizes available information on neurobehavioral asymmetries in animals. Representatives of virtually all main phyla of bilateral invertebrates demonstrate certain degree of neurobehavioral asymmetry (directional or antisymmetry), suggesting that neurobehavioral asymmetries date back to the earliest bilateral organisms and are characteristic for majority of them. Scarcity and irregularity of the studies of neurobehavioral lateralizations in invertebrate species, nevertheless, do not allow us to understand possible homologies and evolutionary continuity of these physiological traits as well as their common or differential gene control. However, one hypothesis has been put forward in that the antisymmetry can be an intermediate state between the symmetry and strong directional asymmetry, i.e., that directional asymmetry is the advanced state of antisymmetry.[11] Phylogenetic studies also indicate that asymmetry could evolve several times from perfect bilateral symmetry.[11,13] On the other hand, asymmetrical neural functions might appear to be shared features in at least some even widely diverged groups, e.g., insects and vertebrates.[1]

Deep routes for both behavioral (brain) and visceral (body) asymmetries (see the previous section of this Chapter) make it difficult to judge on whether these two kinds of asymmetries evolved independently or not. However, there were no examples of clear correspondence of neurobehavioral and situs-related asymmetries in any type or class of invertebrates. Even in the genetically best studied invertebrate model systems like *Caenorhabditis* and *Drosophila* there were no evidence reported so far about any genetic or developmental link between the neural (structural or functional) and visceral (e.g., gut looping) asymmetries. The only animal type, where such a relationship was ever suggested was Chordata. Usually this observation is interpreted as an indicator of a common developmental cause for both kinds of asymmetries. A concerted disappearance of many neurobehavioral and some morphological asymmetries in some amphibian species and their presence in others may partially support this view.[52] However, in the latter review direct links between the situs-related and other asymmetries were not found in any species. Other data also suggest independence of visceral and neural asymmetries, making the possible cause-and-effect relationships mostly illusive.[1,52] It is very likely therefore, that the links between the visceral and brain asymmetries in vertebrates is an evolutionary novelty, being a sort of integration of two independent developmental programs in concerted functioning within complex organism. In the next sections this proposition will be further considered.

Is the Diencephalon Really the Whole Brain?

The existence of neuro-anatomical asymmetries in the diencephalon and their apparent connection to the now well studied Nodal-dependent cascade of genes, leading to asymmetry in visceral organs (especially well described in fish)[53,54] compelled researchers to think that there is only one developmental mechanism responsible for establishment of brain and body asymmetry in ontogeny of vertebrates. However, the lack of any behavioral data, which would connect this asymmetric brain anatomy and visceral situs allowed a different thinking.[1,3] The insufficiency of our knowledge on the role of habenular asymmetry in lateralized behavioral responses (see ref. 55 for a review) and the normal pattern of behavioral handedness in human situs inversus individuals were particular bases of such a point of view.

One difficulty of this view is that some asymmetries in the brain are indeed under the control of situs-related pathway. Such asymmetries include several asymmetric features in the diencephalon. For example, in zebrafish (*Brachydanio rerio*), the neuropil of the left habenular nucleus is larger on the left side, the parapineal organ is situated asymmetrically in the brain to the left of the midline and projecting more neurons to the nearby left habenular nucleus.[53] Initially the structure of the diencephalon develops symmetrically, but at a later stage is becoming lateralized by a stochastic factor or the genes of the Nodal cascade.[56] The functional significance of this anatomical asymmetry was unknown until recently. However, in a very interesting report Barth and colleagues revealed a concordant reversal of laterality in viscera, diencephalon and some behavior (see ref. 57 and commentaries in refs. 58,59) in a progeny of a pair of zebrafish with frequent reversal of asymmetry of the heart, gut and pancreas. In other words, the described *fsi* (*frequent-situs-inversus*) fry demonstrated concordant reversal of organs (in some cases—complete situs inversus phenotype) in contrast to other fish mutants, usually showing heterotaxia. Zebrafish with reversed visceral anatomy had also reversal of epithalamic structure and gene expression pattern. Most interesting that the fry with complete situs inversus also demonstrated reversal of such lateralized behaviors as unilateral viewing of own mirror image and unilateral eye use when approaching the target that the fry were previously trained to bite.[57] However, the fry with situs inversus turned to the same direction as normal individuals when they emerge to the novel environment or when they were startled.

The studies on zebrafish have several interesting key aspects to be further addressed, being at the time uncertain. First, the symmetry condition as the early state in development of epithalamus and perfect establishment of laterality here in the absence of expression in the brain of several known components of the Nodal cascade[53,56] makes it still unlikely that the Nodal cascade is the only genetic pathway, which regulates left-right asymmetry of CNS. It is worth of mentioning here, that even multiple versions of the Nodal-pathway[2,60] acting in the head (epithalamus) and trunk (viscera) are still not enough to support the notion that the cascade is unique in establishing the left-right asymmetry in all parts of the organism. Second, although the connection between the lateralized behaviors and the visceral situs reversal is unequivocal in zebrafish fry, it is still unclear how the diencephalic structures, such as habenular nuclei, are involved in lateralization of viewing behaviors. Is the influence of the Nodal cascade primary or mediated by the physiological (e.g., neural or hormonal) asymmetry in the epithalamus? Interesting, that in chick some (but not all) lateralized behaviors are determined by this cascade indirectly, i.e., via embryonic turning in the egg and induction of neural asymmetry by light stimulation (see Chapter 2 by M. Manns in this volume). Third, the absence of reversion of some lateralized behaviors in zebrafish with situs inversus suggests that these behaviors can be regulated differently from those, which show reversions. Indeed, lateralized monocular viewing, which shows reversion, is a feature of visual processing of information, whereas described kinds of turning, which show no reversions, are almost pure motor behaviors, which have more reflexive nature and dependent on the function of hind regions of the brain and the spinal cord (e.g., Mauthner neuron action in startle response). This grouping of asymmetries reminds that accepted by Malashichev and Wassersug (motor vs. perceptual as showing a certain degree of independence in evolution and development).[1]

Finally, I should stress the importance of knowing the developmental time at which the situs was initially affected in reversed individuals. Either when analyzing the results of behavioral testing of the mutant zebrafish or in the situs inversus humans (see discussion above) we do not know at which developmental time this or that mutation caused the organ reversion. The reversion is probably due to an early event, which should be likely found before neurulation, but how early? In cases of heterotaxic animals or better to say when only certain features of the situs are reversed or when a certain gene is affected, we can sometime know or at least judge on the time of developmental disturbance. Behavioral analyses here would show whether neural features are also reversed or not, thus indicating the establishment time of brain asymmetry. Once we accepting this strategy we should also accept the importance of nonhuman animal model systems other than zebrafish and even may be not traditional ones to address the genetic and developmental origin of brain lateralization.

Searching for Right Genes in Right Places

Considering brain lateralization as a unique characteristic of humans seems to be an anachronism. Indeed, as long as the Bianki's work on brain asymmetry in a variety of experimental subjects from fish to mammals, it is known that the physiological differences between the hemispheres exist in all vertebrate classes.[60-62] A great bulk of recent literature arose on lateralization of aggressive, feeding and other types of behaviors in vertebrates (e.g., fish,[35,64] amphibians,[32-34,41,42,65-67] reptiles,[68,69] birds,[70-74] or nonhuman mammals[75-77] and see also refs. 78,79 and other chapters in the current volume for more examples and further references). Recent advances in the study of invertebrate lateralization (see above) add significantly to the support of deep roots of lateralization, dating it back to early bilateral organisms.

The pattern of vertebrate lateralization is common for many if not all species.[33,80] Even the asymmetry in language related areas in the brain of nonhuman primates is strikingly similar to that found in humans.[81] Moreover, the differences between the human brain and the brain of the relative primates lie probably in the upregulation of many genes belonging to a number of different classes and connected to the elevated level of neuronal activity in the human cortex, rather than in the number or a specific set of genes which would express exclusively in the human brain.[82] Therefore, the difference between the human and a monkey, or even a bird brain may be in some aspects (e.g., brain asymmetry) quantitative rather than qualitative. This knowledge allows us to use for the study of developmental mechanisms of brain asymmetry a broad range of animal species and not necessary only those, which are the established laboratory models.[1] This raises an intriguing possibility to search the handedness and language associated genes in nonhuman species, which possess cerebral cortex (e.g., chick) or even more broadly. This approach may give faster results, especially if we consider difficulties associated with the study of early human development.

Nevertheless, last year a number of asymmetrically expressed genes have been found in the neocortex of the human embryos (relatively old, of age of 12-19 weeks).[83] Although the survey did not uncover the mechanism of early formation of the interhemispheric asymmetry, which should be addressed probably at the time before neurulation,[1] it allowed Sun and colleagues[83] to propose two alternative models of its specification, i.e., via different topographic mapping in the two hemispheres, or via a difference in the tempo of cortical development, with the right hemisphere's development leading over the left.[83] Again, this work also supports the view that genes implicated in visceral asymmetries are not detectable as establishing cerebral asymmetries. The triggering mechanism of brain asymmetry formation in either case remains to be uncovered. It is therefore interesting to consider other models, e.g., de novo generation of asymmetry in originally symmetric CNS after interaction with other persistent asymmetries of the animal body or environment. In this case the differences between the cerebral hemispheres should be likely quantitative in that one will not find here a few strictly asymmetrically expressed genes, like in the Hensen's node (or Spemann's organizer), but the overall level of expression of many genes will be slightly different in the contralateral hemispheres after a given stage of its development (see ref. 83 for comparison). Such a quantitative asymmetry would correspond to the level of neuronal activity and expression changes related to energy metabolism, much like in the case of difference between the human and nonhuman brains just mentioned above (ref. 82, and see also Chapter 14 of this volume by M. Chernisheva for the related scenario).

One question remains on whether the neural lateralization is environmentally triggered or it is still mainly an intrinsic feature of the organism and exists under strict gene controlling mechanisms. Indeed, why the neurobehavioral asymmetries so unstable in populations and sometimes present only at the level of individuals?[1,3,6] A plausible explanation perhaps lies in the variety of species and methods used, as well as plasticity of the neural system, highly dependent on environmental and hormonal modulations during the animal ontogeny and different life periods. It is therefore very difficult to determine lateralization in a species, if an inappropriate method is applied in an inappropriate time of its life cycle. Consider also different locomotion patterns as a possible biasing factor for motor lateralization in frogs and locusts.[40-42] This means that even if a species possesses pronounced functional asymmetry in the neural system (probably all the vertebrates) this may not necessary expresses in particular behavior and will not be certainly revealed in a particular test.

As an alternative explanation, the hypothesis by Vallortigara and Rogers[19] proposes an external mechanism for alignment of the direction of behavioral asymmetries at the population level by means of social interaction between individuals (see above). Interesting, however, that genetically controlled neural circuitry underlies human social behavior.[84] The activation and function of the amygdala in subjects with Williams-Beuren syndrome (WBS), which is characterized by unique hyper sociability combined with increased nonsocial anxiety, is abnormal. Moreover, although not explicitly stated in the cited report, it is suggestive from the presented data, that normal control individuals had functional asymmetry in amygdala with a greater activation on the right side of the brain, while in WBS subjects this asymmetry disappears with a tendency to reversal. If the latter conclusion is valid (see ref. 85 for related supporting data), that would mean that at least in human populations not the social behavior aligns the behavioral asymmetries, but vice versa—the intrinsic, genetically determined asymmetry in the brain may determine the presence of social behavior. A similar fact, when a relatively straightforward genetic mechanism underlies a complex social behavior was once already shown in woles. In these rodents vasopressin V1a receptor expresses at higher level in social and monogamous as compared to more solitary, but polygamous species determining the species differences. Manipulation with the level of expression could dramatically change the species behavior.[86,87]

Conclusion

This chapter was aimed to determine whether asymmetries of the brain and the body are historically and developmentally connected. Although our current knowledge of this is scarce and contradictory, it seems more likely that more than one set of triggering and aligning developmental mechanisms exist, which determine brain and visceral asymmetries in vertebrates. This itself does not exclude their links during embryonic development. Both kinds of asymmetries may have very deep evolutionary roots, dating back to first bilateral organisms, thus making difficult unraveling the tangle of their interactions and evolutions. Fortunately, this also opens the laterality researchers multiple opportunities to choose between a variety of available experimental models.

References

1. Malashichev YB, Wassersug RJ. Left and right in the amphibian world: Which way to develop and where to turn? BioEssays 2004; 26(5):512-522.
2. Levin M. Left-right asymmetry in embryonic development: A comprehensive review. Mech Dev 2005; 122:3-25.
3. Cooke J. Developmental mechanism and evolutionary origin of vertebrate left/right asymmetries. Biol Rev 2004; 79:377-407.
4. Meinhardt H. The radial-symmetric hydra and the evolution of the bilateral body plan: An old body became a young brain. BioEssays 2002; 24(2):185-191.
5. Vitale E, Brancolini V, De Rienzo A et al. Suggestive linkage of situs inversus and other left-right axis anomalies to chromosome 6p. J Med Genet 2001; 38:182-185.
6. Wilson GN. A model for human situs determination. Laterality 1996; 1(4):315-329.
7. Wilson GN, Stout JP, Schneider NR et al. Balanced translocation 12/13 and situs abnormalities: Homology of early pattern formation in man and lower organisms? Am J Med Genet 1992; 38(4):601-607.
8. Trostanetzki MM. Zur Frage über den situs viscerum inversus totalis. Bull l'Inst Lesshaft 1924; 9:7-24.
9. Tanaka S, Kanzaki R, Yoshibayashi M et al. Dichotic listening in patients with situs inversus: Brain asymmetry and situs asymmetry. Neuropsychologia 1999; 37:869-874.
10. Kennedy DN, O'Craven KM, Ticho BS et al. Structural and functional brain asymmetries in human situs inversus totalis. Neurology 1999; 53:1260-1265.
11. Levin M. Twinning and embryonic left-right asymmetry. Laterality 1999; 4(3):197-208.
12. Hobert O, Johnston Jr RJ, Chang S. Left-right asymmetry in the nervous system: The Caenorabditis elegans model. Nature Reviews Neurosci 2002; 3:629-640.
13. Palmer AR. From symmetry to asymmetry: Phylogenetic patterns of asymmetry variation in animals and their evolutionary significance. Proc Natl Acad Sci USA 1996; 93:14279-14286.
14. Nogi T, Yuan YE, Sorocco D et al. Eye regeneration assay reveals an invariant functional left-right asymmetry in the early bilaterian, Dugesia japonica. Laterality 2005; 10(3):193-205.
15. Levin M, Thorlin T, Robinson KR et al. Asymmetries in H^+/K^+-ATPase and cell membrane potentials comprise a very early step in left-right patterning. Cell 2002; 111:77-89.
16. Shimeld SM. Regulation of left-right asymmetry in Ciona. In: Lemaire P, ed. International Urochordate Meeting. France: Carry le Rouet, 2003, (http://nsm.fullerton.edu/~lamberts/ascidian/UromeetingAbstracts.html).
17. Ishii Y, Hibino T, Nishino A et al. An inhibitor for H^+/K^+-ATPase disrupted the left-right asymmetry in the echinoid embryo. Japan: Hakodate, 74th Annual Meeting of the Zoological Society of Japan, 2003.
18. Shimizu H, Fujisawa T. Peduncle of Hydra and the heart of higher organisms share a common ancestral origin. Genesis 2003; 36:182-186.
19. Vallortigara G, Rogers LJ. Survival with an asymmetrical brain: Advantages and disadvantages of cerebral lateralization. Behav Brain Sci 2005; 28(4):575-633.
20. Sovrano VA, Bisazza A, Vallortigara G. Lateralization of response to social stimuli in fishes: A comparison between different methods and species. Psychology and Behavior 2001; 74:237-244.
21. Frisch Kv. The language of bees. Science Progress 1937; 32(125):29-37.
22. Frisch Kv. Die Tanze der Bienen. Österreiche Zoologie Zeitschrift 1946; 1:1-48.
23. Giurfa M. Cognitive neuroethology: Dissecting nonelemental learning in a honeybee brain. Cur Opin Neurobiol 2003; 13:726-735.
24. Giurfa M, Menzel R. Insect visual perception: Complex abilities of simple nervous systems. Curr Opin Neurobiol 1997; 7:505-513.

25. Herridge GA. The honeybee (Apis mellifera) detects bilateral symmetry and discriminates its axis. J Insect Physiol 1996; 42:755-764.
26. Kells AR, Goulson D. Evidence for handedness in bumblebees. J Insect Behav 2001; 14(1):47-55.
27. Udalova GP, Karas AY. Asymmetry of motion direction of Myrmica rubra ants during labyrinth training. Zhournal Vysshei Nervnoi Dejatelnosti 1985; 35(2):377-379, (In Russian).
28. Udalova GP, Karas AY. Asymmetry of motion direction of Myrmica rubra ants during learning in a maze in conditions of alimentary motivation. Zhournal Vysshei Nervnoi Dejatelnosti 1986; 36(4):707-714, (In Russian).
29. Heuts BA, Brunt T. Behavioral left-right asymmetry extends to arthropodes. Behav Brain Sci 2005; 28(4):601-602.
30. Tkhorzhevsky VV. The study of the side choice reaction in mice. Zhournal Vysshei Nervnoi Dejatelnosti 1973; 23(3):659-667, (In Russian).
31. Sherman G, Garbanati G, Rosen G et al. Brain and behavioral asymmetries for spatial preference in rats. Brain Res 1980; 192(1):61-67.
32. Bisazza A, De Santi A, Bonso S et al. Frogs and toads in front of a mirror: Lateralization of response to social stimuli in five tadpole amphibians. Behav Brain Res 2002; 134:417-424.
33. Rogers LJ. Lateralized function in anurans: Comparison to lateralization in other vertebrates. Laterality 2002; 7:219-239.
34. Wassersug RJ, Yamashita M. Assessing and interpreting lateralised behaviors in anuran larvae. Laterality 2002; 7:241-260.
35. Sorvano VA, Rainoldi C, Bisazza A et al. Roots of brain specializations: Preferential left-eye use during mirror-image inspection in six species of teleost fish. Behav Brain Res 1999; 106:175-180.
36. Sreng L. Sensory asymmetries in the olfactory system underlie sexual pheromone communication in the cockroach Nauphoeta cinerea. Neurosci Lett 2003; 351(3):141-144.
37. Pascual A, Huang KL, Neveu J et al. Neuroanatomy: Brain asymmetry and long-term memory. Nature 2004; 427(6975):605-606.
38. Rossi S, Miniussi C, Pasqualetti P et al. Age-related functional changes of prefrontal cortex in long-term memory: A repetitive transcranial magnetic stimulation study. J Neurosci 2004; 24(36):7939-7944.
39. Faure PA, Hoy RR. Auditory symmetry analysis. J Exp Biol 2000; 203:3209-3223.
40. Faisal AA, Matheson T. Coordinated righting behavior in locusts. J Exp Biol 2001; 204:637-648.
41. Robins A, Lippolis G, Bisazza A et al. Lateralized agonistic responses and hind-limb use in toads. Animal Behavior 1998; 56:875-881.
42. Malashichev YB. One-sided limb preference is linked to alternating-limb locomotion in anuran amphibians. J Comp Psychol 2006; 120, (in press).
43. Ades C, Novaes Ramires E. Asymmetry of leg use during prey handling in the spider Scylodes globula (Scytodidae). J Insect Behavior 2002; 15(4):563-570.
44. Udalova GP, Karas AY, Zhukovskaia MI. Asymmetry of the movement direction in Gammarus oceanicus in the open field test. Zhournal Vysshei Nervnoi Dejatelnosti 1990; 40(1):93-101.
45. Byrne RA, Kuba MJ, Meisel DA. Lateralized eye use in Octopus vulgaris shows antisymmetrical distribution. Animal Behavior 2004; 68:1107-1114.
46. Deliagina TG, Orlovsky GN, Selverston AI et al. Asymmetrical effect of GABA on the postural orientation in Clione. J Neurophysiol 2000; 84(3):1673-1676.
47. Reynolds GP, Czudek C, Andrews HB. Dificit and hemispheric asymmetry of GABA uptake sites in the hippocampus in schizophrenia. Biol Psychiatry 1990; 27(9):1038-1044.
48. Chang S, Johnson RJJ, Hobert O. A transcriptional regulatory cascade that controls left/right asymmetry in chemosensory neurons of C. elegans. Genes Dev 2003; 17(17):2123-2137.
49. Hutter H, Wacker I, Schmid C et al. Novel genes controlling ventral cord asymmetry and navigation of pioneer axons in C. elegans. Dev Biol 2005; 284(1):260-272.
50. Shankland M, Martindale MQ. Segmental specificity and lateral asymmetry in the differentiation of developmentally homologous neurons during leech embryogenesis. Dev Biol 1989; 135(2):431-448.
51. Bianki VL, Sheiman IM, Zubina EV. Preference of movement direction in T-maze in Planaria. Zhournal Vysshei Nervnoi Dejatelnosti 1990; 40(1):102-107, (In Russian).
52. Malashichev YB. Asymmetries in amphibians: A review of morphology and behavior. Laterality 2002; 7(3):197-217.
53. Concha ML, Burdine RD, Russell C et al. A Nodal signaling pathway regulates the laterality of neuroanatomical asymmetries in the zebrafish forebrain. Neuron 2000; 28:399-409.
54. Concha ML, Wilson SW. Asymmetry in the epithalamus of vertebrates. J Anat 2001; 199:63-84.
55. Vallortigara G, Rogers LJ, Bisazza A. Possible evolutionary origins of cognitive brain lateralization. Brain Res Rev 1999; 30:164-175.

56. Concha ML, Russell C, Regan JC et al. Local tissue interactions across the dorsal midline of the forebrain establish CNS laterality. Neuron 2003; 39:423-438.
57. Barth KA, Miklosi A, Watkins J et al. fsi zebrafish show concordant reversal of laterality of viscera, neuroanatomy, and a subset of behavioral responses. Current Biology 2005; 15:844-850.
58. McManus C. Reversed bodies, reversed brains, and (some) reversed behaviors: Of zebrafish and men. Dev Cell 2005; 8:796-797.
59. Craven R. Mirror images of asymmetry. Nature Rev Neurosci 2005; 6:663.
60. Bisgrove BW, Essner JJ, Yost HJ. Multiple pathways in the midline regulate concordant brain, heart and gut left-right asymmetry. Development 2000; 127:3567-3579.
61. Bianki VL. On ontogenesis of the paired function of [brain] hemispheres in caudate amphibians. Biologicheskije nauki 1967; 6:53-55.
62. Bianki VL. Asymmetry of the animal brain. Leningrad: Nauka, 1985.
63. Bianki VL. Asymmetry of the brain as the basis of the animal behavior. Proceedings of the Biological Institute Leningrad State University 1990; 41:138-152.
64. Bisazza A, Cantalupo C, Capocchiano M et al. Population lateralisation and social behavior: A study with 16 species of fish. Laterality 2000; 5:269-284.
65. Green A. Asymmetrical turning during spermatophore transfer in the male smooth newt. Animal Behavior 1997; 54(2):343-348.
66. Bisazza A, Cantalupo C, Robins A et al. Pawedness and motor asymmetries in toads. Laterality 1997; 2:49-64.
67. Wassersug RJ, Naitoh T, Yamashita M. Turning bias in tadpoles. Journal of Herpetology 1999; 33:543-548.
68. Deckel AW. Laterality of aggressive responses in Anolis. J Exp Zool 1995; 272:194-200.
69. Robins A, Chen P, Beazley LD et al. Lateralized predatory responses in the Ornate dragon lizard (Ctenophorus ornatus). Neuroreport 2005; 16(8):849-852.
70. Andrew RJ. The development of visual lateralization in the domestic chick. Behav Brain Res 1988; 29:201-209.
71. Rogers LJ, Workman L. Footedness in birds. Anim Behav 1993; 45:409-411.
72. Diekamp B, Prior H, Güntürkün O. Lateralization of serial color reversal learning in pigeons (Columbia livia). Animal Cogn 1999; 2:187-196.
73. Diekamp B, Regolin L, Güntürkün O et al. A left-sided visuospatial bias in birds. Current Biology 2005; 15:R372-R373.
74. Csermely D. Lateralization in birds of prey: Adaptive and phylogenetic considerations. Behavioral Processes 2004; 67:511-520.
75. Denenberg VH, Garbanati J, Sherman GF et al. Infantile stimulation induces brain lateralization in rats. Science 1978; 301:1150-1152.
76. Westergaard GC, Suomi SJ. Hand preference for a bimanual task in tufted capuchins (Cebus apella) and rhesus macaques (Macaca mulatta). J Comp Psychol 1996; 110:406-411.
77. Yaman S, Fersen L, Dehnhardt G et al. Visual lateralization in the bottlenose dolphin (Tursiops truncatus): Evidence for a population asymmetry? Behav Brain Res 2003; 142:109-114.
78. In: Andrew RJ, Rogers LJ, eds. Comparative Vertebrate Lateralization. Cambridge: Cambridge University Press, 2002.
79. In: Malashichev YB, Rogers LJ, eds. Behavioral and Morphological Asymmetries in Amphibians and Reptiles. Laterality (Special Issue) 2002; 7(3):1-96.
80. Rogers LJ. Lateralization in vertebrates: Its early evolution, general pattern, and development. Advances in the study of behavior 2002; 31:107-161.
81. Cantalupo C, Hopkins WD. Asymmetric Broca's area in great apes. Nature 2001; 414:505.
82. Caceres M, Lachuer J, Zapala MA et al. Elevated gene expression levels distinguish human from nonhuman primate brains. PNAS 2003; 100(22):13030-13035.
83. Sun T, Patoine C, Abu-Khalil A et al. Early asymmetry of gene transcription in embryonic human left and right cerebral cortex. Science 2005; 308:1794-1798.
84. Meyer-Lindenberg A, Hariri AR, Munoz KE et al. Neural correlates of genetically abnormal social cognition in Williams syndrome. Nature Neurosci 2005; 8(8):991-993.
85. Noesselt T, Driver J, Heinze HJ et al. Asymmetrical activation in the human brain during processing of fearful faces. Current Biology 2005; 15:424-429.
86. Lim MM, Wang Z, Olazabal DE et al. Enhanced partner preference in a promiscuous species by manipulating the expression of a single gene. Nature 2004; 429(6993):754-757.
87. Pitkow LJ, Sharer CA, Ren X et al. Facilitation of affiliation and pair-bond formation by vasopressin receptor gene transfer into the ventral forebrain of a monogamous vole. J Neurosci 2001; 21(18):7392-7396.

SECTION II
Eye Use and Cerebral Lateralization

CHAPTER 5

An Eye for a Predator:
Lateralization in Birds, with Particular Reference to the Australian Magpie

Lesley Rogers* and Gisela Kaplan

Abstract

Avian species with their eyes placed laterally on the sides of their head show eye preferences for viewing stimuli at a distance, as determined by the angle of the head adopted when they use the monocular field of vision. Studies of a number of species have revealed that eye preferences are present at the level of the population. Here we were most interested in discussing an apparently general pattern for the left eye to be used to view novel stimuli and stimuli demanding detection and rapid response, as in the case of responding to a predator. We discuss the evidence for this in the domestic chick and some other avian species and then consider lateralized eye use in the Australian magpie tested in its natural environment. We report our recent finding that playback of a specific "eagle" alarm call to magpies elicits looking up with the left eye and contrast this with the absence of eye/ear preferences in magpies during foraging. We also report that magpies use their left eye to track and locate moving food objects (equivalent to insects). We conclude that magpies have the same pattern of lateralization shown previously in laboratory studies of the domestic chick and we discuss the structural asymmetry of the visual pathways and relate the eye preferences to differences between the hemispheres for processing visual information.

Hemispheric Specialization and Eye Preferences in Birds

The hemispheres of the avian brain are specialized to process perceptual inputs in different ways and to control different motor functions. We know this from a large number of studies of the domestic chick and the pigeon (summarized in ref. 1), but few studies have investigated lateralization of avian species in their natural habitat. This chapter reports on preferred use of the left eye to examine novel stimuli and predators in laboratory studies of the young domestic chick (*Gallus gallus domesticus*) and field studies of the Australian magpie (*Gymnorhina tibicen*).

Eye preferences to view stimuli can be determined quite easily by measuring the angle of the head adopted by the bird when it fixates the stimulus (see method in ref. 2). A number of studies have shown that the domestic chick prefers to view familiar and novel stimuli using different eyes. One study, by McKenzie et al[3] (1998), found that, once chicks have imprinted on a model social partner, they use the left eye (LE) when viewing the familiar social partner at a distance, and, when they have not imprinted, they use the right eye (RE) when they look at an attractive stimulus on which they might imprint (e.g., a hen). While this might suggest that the LE is used for monitoring familiar stimuli and the RE for attending to attractive novel

*Corresponding Author: Lesley Rogers—Centre for Neuroscience and Animal Behavior, Building W28, University of New England, Armidale, NSW 2350, Australia. Email: lrogers@une.edu.au

Behavioral and Morphological Asymmetries in Vertebrates, edited by Yegor B. Malashichev and A. Wallace Deckel. ©2006 Landes Bioscience.

stimuli, this simple conclusion does not seem to be correct since other experiments have shown that use of the LE is more likely to lead to a response to novel stimuli than use of the RE.[4,5] Hence, the RE might have a specific role to attend to an imprinting stimulus, rather than to novelty. The RE is also preferred when the chick approaches a stimulus that it must manipulate, such as a bowl with lid that the chick must remove in order to obtain a food reward.[6,7] The LE is used when the chick approaches a food bowl without a lid.

Predator detection, as well as further examination of predators, is also carried out by use of the LE. A chick, engaged in foraging detects a predator advancing overhead sooner when it approaches on its left side than when it advances on the right.[8] It also prefers to examine (i.e., continue viewing) the overhead predator using its left eye.[9,10] In fact, Dharmaretnam and Rogers[9] found that, if a chick detected the overhead predator with its RE, it would turn its head to examine the predator further using its LE.

The specialization of the LE and right hemisphere for responding to predators is further supported by an earlier report by Evans et al[11] showing that adult domestic chickens scan overhead using the LE when they hear their species-typical call signalling the presence of an aerial predator. This response is entirely consistent with preferential use of the LE in response to novelty, and hence predator detection, and it is also consistent with other research showing that the chick uses its LE when it attends to global, spatial cues. Tommasi et al[12] trained young domestic chicks to find food buried under sawdust in the centre of an arena and marked by a small landmark (a rod with a small flag) as a local cue. Then they tested the chick's searching behavior when it was placed in the arena, now with no food buried in the centre and the landmark displaced to a position nearer one of the walls. The chicks were tested monocularly using either the LE or RE. Chicks using the LE searched in the centre of the arena, which meant that they were relying on geometric, spatial cues, whereas those tested using the RE used local cues (i.e., they searched near the landmark; see also ref. 13).

The overall picture that is emerging is that the LE-right hemisphere is specialized for a constellation of functions dependent on use of global cues and essential for locating predators and food. However, this specialization might extend to other situations in which rapid decisions and appropriate responses have to be made. One of these is in agonistic social interactions. In fact, the chick is more likely to precede aggressive pecking of an unfamiliar chick by monocular viewing with the left eye rather than by the right eye.[14]

We might also ask whether the LE specialization for detection and response might apply to specific stimuli not likely to be seen as potential predators. To investigate this we reexamined some data collected earlier (by Rogers) in a task designed to measure the responses of 8-day-old chicks to small, red beads advanced toward the chick from behind and simultaneously into its left and right monocular visual fields. The apparatus consisted of two beads attached to fine metal rods projecting through slits in the floor. The beads were advanced slowly towards the chick at its eye level and on its left and right side simultaneously while the chick, previously deprived of food, was pecking at a dish of food. Once the chick detected the bead(s) it stopped feeding, looked up and then turned to peck one of the beads. The latter response showed that they did not see the stimulus as a threat, and so it had little, if anything, in common with the tasks testing predator detection. We were interested in which direction the chick turned on the first presentation of the beads, and we found that 18 chicks turned to the bead seen in the monocular field of their LE and only 9 chicks turned to the one seen by their RE. Hence, it seems that the LE advantage for detection and response extends to a range of stimuli.

Consistent with the known functions of the LE and right hemisphere system in the chick, Bugnyar et al[15] noted that hand-raised ravens (*Corvus corax*) would follow the direction of eye gaze of a human experimenter looking upwards and that the birds did so by tilting their head to look up using the monocular field of the LE. This response might well be an expression of the predator-detection system.

Overall, therefore, a number of experimental results obtained by testing birds in the laboratory suggest that the LE-right hemisphere plays a specialized role in visual detection of predators. Hence, we considered it timely to extend this research into the field to see whether an avian species in its natural habitat expressed the same lateralized preference.

Organization of the Visual Pathways in Birds

Preferred use of one eye is usually said to indicate that the opposite hemisphere is being used since the optic fibres cross over completely. However, the hemisphere on the same side as the preferred eye may play some (minor) role since some projections recross the midline in both the tectofugal and thalamofugal visual systems. Added to this we need to take into account that there are asymmetries in the visual projections to the forebrain: in the thalamofugal visual system of the chick[16,17] and the tectofugal system of the pigeon.[18]

As illustrated in Figure 1, in the chick, there are more projections from the left side of the thalamus (which receives input from the right eye, RE) to the right side of the forebrain than there are from the right side of the thalamus (which receives input from the left eye, LE) to the left side of the forebrain.[16,17] Both sides of the thalamus have large numbers of projections that project to the forebrain without crossing the midline of the brain (i.e., to their respective ipsilateral forebrain hemispheres).[16] Overall, therefore, inputs from the LE of the chick are processed almost exclusively by the right hemisphere, whereas inputs from the RE are processed by both the left and right hemispheres, even though the left hemisphere plays a somewhat greater role. Hence, when a chick uses its LE to view a stimulus, and as far as processing in its thalamofugal visual system is concerned, it is opting to process the information in a strongly lateralized manner, by

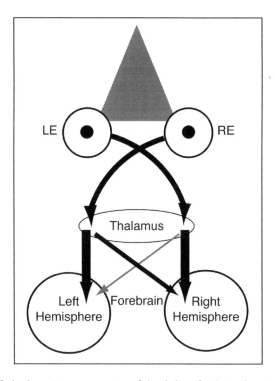

Figure 1. A simplified schematic representation of the thalamofugal visual projections in the chick, showing that there are more projections from the left side of the thalamus to the right forebrain hemisphere (pallium) than from the right side of the thalamus to the left forebrain hemisphere. LE: left eye; RE: right eye. The view presented is as if one were looking downward at the dorsal surface of the brain. Note that the diagram is not to scale.

the right hemisphere. Use of its RE may engage the left hemisphere primarily, but not exclusively. In other words, in terms of the thalamofugal system only, preferred viewing of stimuli by the monocular field of the LE may reflect more strongly lateralized use of the forebrain than does RE viewing. We note however, that this does not apply to the tectofugal system, which is not lateralized in the chick.[16] In the tectofugal system, the main input from each eye is transmitted to the contralateral hemisphere. Since the tectofugal system has no asymmetry in the chick it does not concern us here. Nevertheless, stimuli detected and viewed by the LE would be processed almost exclusively by the right hemisphere irrespective of whether the thalamofugal or tectofugal systems are involved. We conclude, therefore, that the right hemisphere is specialized for detecting novel stimuli, detecting and examining potential predators.

Visual Lateralization in Wild Birds

Eye preferences should be relatively easy to measure in wild birds, at least in those species with their eyes positioned laterally on the sides of their head so that they have large monocular visual fields and only a small binocular field. Characteristically, these species turn the head to view stimuli located at a distance so that they use the lateral, monocular field of vision. They do this because the monocular field is focused at a much greater distance then the myopic binocular field. In fact, some species (e.g., the pigeon) have two foveae, one receiving input from the frontal field and the other from the lateral, monocular field.

To our knowledge only three published studies have reported eye preferences shown by birds in their natural habitat. The first, by Franklin and Lima,[19] recorded the behavior of juncos, *Junco hyemalis*, feeding alongside a barrier so that they had to orient themselves to monitor for predators. The birds showed a preference to orient their left sides next to the barrier and so look outwards with a RE preference. Tree sparrows, *Spizella arborea*, showed no significant preference but a trend in the opposite direction to that of the juncos.

The second study, conducted by us, showed that kookaburras, *Dacelo gigas*, show a strong LE preference to scan for moving prey at a distance (i.e., when the bird is perched on a power line and looking down to the ground directly below it: see ref. 20). A total of 88 birds were scored, once only per bird, and 83% of the scores for looking down involved turning the head to use the monocular visual field. Of these monocular sightings, 86% were LE (63 LE, 10 RE; $z = 6.20$, $p < 0.001$).

The third study, by Ventolini et al,[21] showed that black-winged stilts, *Himantopus himantopus*, have a population-level preference to tilt their head to use the monocular field of the RE before they peck at prey and that such pecks at prey are more successful than are ones using the LE. This study also found that the male birds were more likely to direct courtship displays to females seen in the monocular visual field of their LE than to those seen by the RE. This complementary specialization of the LE and RE systems is entirely consistent with that shown in the domestic chick.

Since it is of interest to extend the investigation of lateralization in birds to a wider range of species and to natural behavior, we are now measuring eye preferences in the Australian magpie, *Gymnorhina tibicen*, as part of a broader study of their communication behavior. We know already that magpies have a large repertoire of alarm calls (twelve categories plus variations) and an extraordinarily large song repertoire. Moreover, magpies are territorial and vigilance behavior extends beyond nest protection to protection of group members and territorial borders[22] so that predator vigilance is highly developed. We further know that magpie alarm calls serve heterospecific species within the same territory.[23]

Moreover, Australian magpies have laterally placed eyes and large monocular visual fields. Using an ophthalmoscope, we have measured the extent of the binocular and monocular visual fields in the horizontal plane passing through the eyes at beak level (Fig. 2). Each eye of the magpie has a monocular field of 143-149° and the binocular field is only 28-34°. This means that the species has about the same sized binocular field as the chicken and the monocular fields are large. Head turning to view stimuli is seen commonly in their activities of foraging (note that this may involve use of auditory cues monaurally, as well as visual cues) and monitoring for predators (Fig. 3). Such monocular viewing can be recorded with relative ease by observing wild magpies through binoculars, as well as by making video recordings.

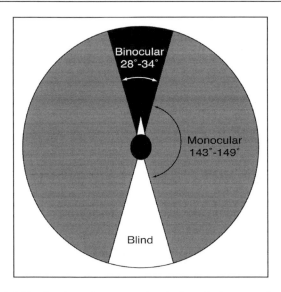

Figure 2. The visual fields of an Australian magpie in a horizontal plane at eye level. The angles were determined, as accurately as possible, using a live hand-raised magpie and gently holding the head at the edges of a small table while another experimenter determined the angles at which the eye could no longer be seen, using an ophthalmoscope. The black area indicates the binocular field, the grey the monocular fields and the white the blind area behind the bird's head. Results from four birds.

Figure 3. Photograph of a wild magpie viewing a raptor overhead. Note the head tilt to allow monocular viewing. Note also that the birds recorded during our tests involving playback of alarm calls were all on the ground at the time of testing.

Playback of Alarm Calls

First we identified alarm calls used by magpies when spotting a bird of prey, particularly an eagle or a goshawk. These calls were recorded in addition to a range of other alarm calls. Although we know that magpie alarm calls show some variations across different geographical areas,[24] one type of alarm call has been found across many different research sites and geographical regions and appears to be widely used. We have termed this the 'generic' alarm call. Generic alarm calls are those used for general, nonpredator related situations by magpies across a wide range of territories in Australia. The 'eagle' alarm call was emitted only when the magpies had seen an eagle (either a living wedge-tailed eagle or taxidermic model of one). The two alarm calls differ markedly in vocal structure. Generic alarm calls have left to right downward sloping formants (i.e., descending frequency) with the dominant formant descending from around 2.5 kHz to 1.8 kHz, giving the sound spectrogram the appearance of twisted columns, whereas eagle alarm calls are broadband, high-pitched and complex, having a more chaotic structure.[22] The details of the structure of these calls (Kaplan and Rogers in preparation) are not relevant to the points that we make here about eye preferences.

'Eagle' alarm calls and generic calls were recorded for separate playback sessions and some of the eagle alarm call variations were randomized for playback experiments. The playback method consisted of a pretest period, a testing period and a post-test period (each of five minutes duration). We monitored the birds' responses to their species-typical alarm calls across all three periods. The playback was conducted using a field amplifier, at ground level, reproducing calls at 70-80 dB at 10 m from the sound source. A control call was included and this was the warble call of the magpie. Hence, we played back four types of recorded calls: generic alarm calls, 'eagle' alarm call, generic plus 'eagle' alarm calls and warbles. The order of presentation was random.

We recorded the head position (and hence eye used) of a focal magpie before, during and after each presentation. Only one magpie was scored per session. Magpies at 10 localities were tested and the presentations of each of four different calls were repeated three times at each location.

Here we present our results for monocular looking overhead. Typically, the bird raises its beak so that it points upwards and it then tilts its head sideways to view overhead with the LE or RE. Looking up without head tilting also occurs, presumably to view the horizon and nearby trees or, less likely, to look overhead at a close focal distance using the binocular field. We recorded these head positions also, but here we are interested only in the eye preferences during monocular looking overhead. Although this head tilting would direct the left ear as well as the left eye upwards, our field observations indicate that the bird is using its eye rather than ear since this head posture is used to view eagles circling very high up in the sky. Auditory cues would not be available in such cases. Nevertheless, use of the left ear, as for the LE, would also mean preferential processing by the right hemisphere and be entirely consistent with our interpretation of the lateralization of visual processing.

Scanning (head turning from side to side) both with the beak up and without looking up (i.e., head held with beak horizontal) also occurs, but it is not of interest here since it is not possible to determine eye preferences, even if they are part of this response. Moreover, the increase in these scores during playback is much less than that recorded for looking up and it is not differentiated according to any of the four calls played back.

Vigilance for Overhead Predators

On hearing the playbacks the magpies ceased any on-going behavior, such as foraging, walking or running. On hearing the 'eagle' alarm call, they looked up to monitor overhead. Slight increases in looking up also occurred when they heard the generic alarm call and the generic plus 'eagle' but these increases were much less than during playback of the 'eagle' alarm call, and the 'looking up' scores did not increase on playback of the warble call (Kaplan and Rogers, in preparation). It is highly important that looking up increases when the magpies hear the 'eagle' alarm call since the magpies were on the ground at the time of testing and the calls were played back via a speaker at ground level. Hence, they made a meaningful response to the call heard; viz., looking overhead for the potential presence of an aerial predator.

We defined looking up as visual fixation with the beak directed upward at least 45° to the horizontal. Whenever the focal bird looked up and showed a clear tilting of the head to look overhead with one eye, this was recorded. Looking up events in which it was not possible to score the angle of the head accurately were recorded but not analysed as part of the data set reported here. Looking up that involved no tilting of the head was also recorded and labelled as 'both eyes' (BE). BE may not have been limited to use of the binocular field because it could have entailed panoramic viewing using the monocular fields as well as the binocular field.

During playback of the eagle alarm call 67 events of looking up could be defined as clearly LE, RE or BE. Note that, although we presented the eagle alarm call many times and recorded many more looking up events, we could not assign more than this number of looking events to one of these three categories with absolute certainty. Of these 67 events, 51 events (76%) were monocular, showing that the lateral visual field is preferred for distant viewing when scanning overhead. In fact, in addition to these scores, we have recorded several events in which a magpie uses the monocular field to view a real eagle circling overhead. Of the monocular events of looking up during playback of the eagle alarm call, 43 events (84%) were LE (binomial z score = 2.32, $p < 0.05$). Hence, the magpie shows preferential use of the LE to scan overhead, and this result is entirely consistent with the laboratory studies of the domestic chick, discussed above. Since these data were collected from birds in a range of very different natural environments, it is most unlikely that the lateral bias was determined by left/right differences in the location of visual or auditory stimuli. Moreover, this is a population bias not likely to be influenced by repeated measures from certain individuals since only one bird in a given locality was scored on each test and rarely was it possible to score more than 1 to 3 LE or RE events of looking up per bird.

To determine whether this LE preference was specific for predator detection, we compared the result to scores for looking up during the experimental period prior to playback of the alarm calls and at other times when no playback occurred (N = 145). These scores were collected over a much longer time span since overhead scanning occurred infrequently when the eagle alarm call was not played. Only 84 (58%) of these looking up events were monocular and there was no significant eye preference (40 LE versus 44 RE, 48% LE).

Too few recordings of looking up were recorded during playback of the generic alarm call and a warble call (not an alarm call) to determine eye preference with any certainty. However, for our sample size of 21 looking up responses to generic alarm calls (each score for a different bird) 8 scores were LE, 7 RE and 6 BE. These preliminary data suggest that, when overhead viewing occurs in response to hearing the generic alarm call, use of the monocular field is more common than use of both eyes, but that there is no preferred eye in this case (see Table 1).

Table 1. Monocular viewing

Behavior	N	% M	% L	Z-Score	P
Looking up on playback of eagle alarm call	67	76	84	2.32	<0.05
Looking up on playback of generic alarm call	21	71	53	0.23	NS
Looking up without call playback	145	58	48	0.44	NS
Looks during foraging	917	29	44[#]	-2.08	<0.05
Viewing moving food	159	97	97	11.65	<0.001

N: total number of 'looks' recorded (sample time varies); % M: percent of looks that were monocular; % L: percent of monocular looks with the left eye; NS: indicates no significant bias; #: this score is likely to refer to ear rather than eye use.

Hence, the 'eagle' alarm call elicits a specific and appropriate response by the receiver (i.e., monitoring at a distance overhead). The specificity of the response is even greater when we consider the preferential use of the LE, and therefore specialized use of the right hemisphere.

It is worth mentioning here the results of another study conducted in our laboratory by Adam Hoffman (submitted), who tested eye preference and fear responses in magpies by approaching them from behind. This elicits head turning to view the experimenter using the monocular field of the LE or RE, usually followed by vocalization and flying away. A total of 34 birds were tested repeatedly so that a laterality index (LE-RE/LE+RE) could be determined for each bird. These scores were correlated with the number of generic alarm calls made by each individual as it took flight and the result was significant (r = 0.67, p = 0.013). The greater the preference to use the LE, the more alarm calls elicited. In other words, bringing the LE and right hemisphere to bear on the task is associated with higher levels of alarm or, at least, alarm calling.

Head Turning during Foraging

Magpies frequently feed on insect larvae of the scarab beetle buried in the ground, which they detect using auditory cues only.[25] They also eat earthworms and other small, ground-dwelling animals. Before probing the ground with its beak, the magpie frequently tilts its head to the left or right, presumably to listen with one or the other ear.[23] Although visual cues might be used also, it is unlikely that the head turning behavior while foraging relies entirely on visual mechanisms since the part of the lateral, monocular field used would not be focused at the short distance of ground level, whereas the frontal, binocular field is focussed for this short distance. Hence, visual inputs guiding pecking would be likely to depend on the frontal field rather than the lateral, monocular field. In fact, a binocular sighting follows the head turning and precedes delivery of a probing peck (i.e., a peck that penetrates beneath the surface).

We have scored the direction of head turning for approximately 20 magpies foraging on the ground in six different locations. A total of 917 pecking events were recorded, 266 (29%) of which were preceded by turning of the head. Of these, 116 (44%) were cases of turning the left ear (and eye) towards the ground (z = -2.08, p < 0.05). Hence, there was a slight but significant bias at population level for the bird to turn its head so that the right eye and ear were directed towards the ground. This result is consistent with preferred use of the left hemisphere (and right eye) in control of feeding responses, as demonstrated previously in the chick,[26] pigeon[18] and zebra finch,[27] although the bias for use of the RE is stronger in these latter species than in the magpie. The difference in strength of bias could be a species difference or, as we believe more likely, it could be due to the use of auditory cues by the magpie and not the other species, the latter being grain-eaters.

This opposite direction of head turning during foraging versus predator detection shows that the magpie has no motor bias to turn the head only rightward or only leftward and, therefore, that the left eye preference to monitor overhead for an eagle is a perceptual bias, and not merely secondary to a motor bias.

Eye Preference for Tracking Moving Prey

Magpies living in the vicinity of humans respond readily to food provisioning. They soon learn to take small pieces of raw, minced meat thrown to them and are skilful in catching them in the air or as soon as they land on the ground. This tendency to befriend people and to live near humans can be effectively exploited for research in the field. Magpies will come within 2 m of the food being provided, thus affording easy and accurate observation of eye use just prior to retrieval of the food reward.

We tested 12 magpies and scored the eye used to track small pieces of meat thrown to them by a human. Individual birds were not identified. A total of 159 events were scored and

155 (97%) of these involved monocular viewing before retrieval. In the latter, a strong left eye preference (97% LE) was found (150 LE, 5 RE; z = 11.65, p < 0.001). This LE preference would seem to reflect specialization of the right hemisphere for processing spatial information using global cues and for controlling rapid responses, as shown in the domestic chick (spatial processing;[13,28] rapid responding[29]).

Although the testing procedure we used here for the magpies was artificial, we have observed use of the LE by magpies feeding naturally on locusts. These observations were made at Kallara, far western NSW, Australia, during a plague of locusts. Two juvenile magpies spent a considerable amount of time attempting to capture the locusts that took flight from the ground as the magpie ran along. Although the attempts were only occasionally successful, the magpie typically tilted its head and even part of its body so as to angle the LE forward while, at the same time, it ran forward and attempted to grasp in its beak an insect flying at eye height.

These particular results suggest that in the magpie, as in the domestic chick, the LE is used preferentially to detect a range of moving targets, food and predators, to which the bird has to respond rapidly.

Conclusions

Our research on Australian magpies not only demonstrates lateralization of visual behavior in an avian species in its natural habitat but also, we believe, provides a methodology by which the study of lateralization can be extended to fieldwork. Furthermore, by taking lateralized responses into account in the field we have enhanced the power of our observations investigating referential signalling in the Australian magpie.

It now seems clear that the domestic chick and the magpie have remarkably similar hemispheric specializations. These are manifested as preferred use of the monocular field of the left eye, and right hemisphere, to detect and control responses to predators and to attend to spatial cues in order to track moving objects or insects before grasping them. As described in the introduction, inputs from the LE are processed by the right hemisphere almost exclusively, and in a more lateralized way than inputs to the forebrain from the RE. We suggest that the strong lateralization of LE inputs underlies the role of this eye system in controlling rapid responses essential for survival. A need to process information by neural circuits in both hemispheres may be advantageous for some tasks but not ones demanding rapid decisions and responses.

As we know from studies of the chick (summarized in ref. 30), the RE system is specialized for guiding pecking for food grains that must be discriminated visually from the background and for guiding responses that require manipulation of objects (e.g., removal of a lid from a food bowl, as discussed above), as well as to attend to local landmarks and other local cues and to inhibit agonistic responses.

This pattern of complementary hemispheric specialization seen in avian species has many similarities to the pattern seen in fish,[31,32] reptiles,[33,34] amphibians[35-37] and mammals.[30,38] The commonality suggests that having a lateralized brain is important for survival and, indeed, some recent evidence shows that lateralized chicks are able to perform more than one task simultaneously much better than can chicks that have no lateralization for these particular tasks.[10] Taken together with the other studies showing lateralization of eye use in avian species in their natural environment, our findings indicate that side biases in behavior are not simply an artefact of testing animals in the laboratory but a quite common aspect of natural behavior.

Acknowledgements

We are grateful to the Australian Research Council for funding this research. We also thank Dr. M. Sawyer and Ms. Jane Hall for assistance in collecting some of the scores on eye use in response to playback of the calls.

References

1. Rogers LJ, Andrew RJ, eds. Comparative Vertebrate Lateralization. Cambridge: Cambridge University Press, 2002.
2. Dharmaretnam M, Andrew RJ. Age- and stimulus-specific use of right and left eyes by the domestic chick. Anim Behav 1994; 48:1395-1406.
3. McKenzie R, Andrew RJ, Jones RB. Lateralization in chicks and hens: New evidence for control of response by the right eye system. Neuropsychologia 1998; 36: 51-58.
4. Andrew RJ. The nature of behavioral lateralization in the chick. In: Andrew RJ, ed. Neural and Behavioral Plasticity: The Use of the Domestic Chick as a Model. Oxford: Oxford University Press, 1991:536-554.
5. Vallortigara G, Andrew RJ. Lateralization of response to change in social partner in chick. Anim Behav 1991; 41:187-194.
6. Andrew RJ, Tommasi L, Ford N. Motor control by vision and the evolution of cerebral lateralization. Brain Lang 2000; 73:220-235.
7. Tommasi L, Andrew RJ. The use of viewing posture to control visual processing by lateralized mechanisms. J Exp Biol 2002; 205:1451-1457.
8. Rogers LJ. Evolution of hemispheric specialisation: Advantages and disadvantages. Brain Lang 2000; 73:236-253.
9. Dharmaretnam M, Rogers LJ. Hemispheric specialization and dual processing in strongly versus weakly lateralized chicks. Behav Brain Res 2005; 162:62-70.
10. Rogers LJ, Zucca P, Vallortigara G. Advantage of having a lateralized brain. Proc Roy Soc Lond B 2004; 271:S420-S422.
11. Evans CS, Evans L, Marler P. On the meaning of alarm calls: Functional references in an avian vocal system. Anim Behav 1993; 46:23-28.
12. Tommasi L, Gagliardo A, Andrew RJ et al. Separate processing mechanisms for encoding geometric and landmark information in the avian hippocampus. Europ J Neurosci 2003; 17:1695-1702.
13. Tommasi L, Vallortigara G. Encoding of geometric and landmark information in the left and right hemispheres of the avian brain. Behav Neurosci 2001; 115:602-613.
14. Vallortigara G, Cozzutti C, Tommasi L et al. How birds use their eyes: Opposite left-right specialisation for the lateral and frontal visual hemifield in the domestic chick. Curr Biol 2001; 11:29-33.
15. Bugnyar T, Stöwe M, Heinrich B. Ravens, Corvus corax, follow gaze direction of humans around obstacle. Proceed Roy Soc Lond B 2004; 271:1331-1336.
16. Rogers LJ, Deng C. Light experience and lateralization of the two visual pathways in the chick. Behav Brain Res 1999; 98:277-287.
17. Rogers LJ, Sink HS. Transient asymmetry in the projections of the rostral thalamus to the visual hyperstriatum of the chicken, and reversal of its direction by light exposure. Exp Brain Res 1988; 7:378-384.
18. Güntürkün O. Ontogeny of visual asymmetry in pigeons. In: Rogers LJ, Andrew RJ, eds. Comparative Vertebrate Lateralization. Cambridge: Cambridge University Press, 2002:247-273.
19. Franklin IIIrd WE, Lima SL. Laterality in avian vigilance: Do sparrows have a favourite eye? Anim Behav 2001; 62:879-885.
20. Rogers LJ. Advantages and disadvantages of lateralization. In: Rogers LJ, Andrew RJ, eds. Comparative Vertebrate Lateralization. Cambridge: Cambridge University Press, 2002:126-153.
21. Ventolini N, Ferrero E, Sponza S et al. Laterality in the wild: Preferential hemifield use during predatory and sexual behavior in the Black winged stilt (Himantopus himantopus). Anim Behav 2005; 69:1077-1084.
22. Kaplan G. Alarm calls, communication and cognition in Australian magpies, symposium paper, International Ornithology Congress, Beijing, 2002. Acta Zoologica Sinica 2005; In press.
23. Kaplan G. Australian magpie: Biology and behavior of an unusual songbird. Collinwood: CSIRO Publishing/UNSW Press, 2004.
24. Brown ED, Farabaugh SM. Macrogeographic variation in alarm calls of the Australian magpie Gymnorhina tibicen. Bird Behav 1991; 9:64-68.
25. Floyd RB, Woodland DJ. Localization of soil dwelling scarab larvae by the black-backed magpie, Gymnorhina tibicen (Latham). Anim Behav 1981; 29:510-17.
26. Rogers LJ. Early experiential effects on laterality: Research on chicks has relevance to other species. Laterality 1997; 2:199-219.
27. Alonso Y. Lateralization of visually guided behavior during feeding in zebra finches (Taeniopygia guttata). Behav Proc 1998; 43:257-263.

28. Tommasi L, Vallortigara G. Hemispheric processing of landmark and geometric information in male and female domestic chicks (Gallus gallus). Behav Brain Res 2004; 155:85-96.
29. Andrew RJ, Rogers LJ. The nature of lateralization in tetrapods. In: Rogers LJ, Andrew RJ, eds. Comparative Vertebrate Lateralization. Cambridge: Cambridge University Press, 2002:94-125.
30. Rogers LJ. Lateralization in vertebrates: Its early evolution, general pattern and development. In: Slater PJB, Rosenblatt J, Snowdon C et al, eds. Advances in the Study of Behavior. 2002:31:107-162.
31. Cantalupo C, Bisazza A, Vallortigara G. Lateralization of predator-evasion response in a teleost fish (Girardinus falcatus). Neuropsychologia 1995; 33:1637-1646.
32. Miklósi A, Andrew RJ, Savage H. Behavioral lateralization of the tetrapod type in the zebrafish (Brachydanio rerio), as revealed by viewing patterns. Physiol Behav 1998; 63:127-135.
33. Deckel AW. Laterality of aggressive responses in Anolis. J Exp Zool 1995; 272:194-200.
34. Hews DK, Worthington RA. Fighting from the right side of the brain: Left visual field preference during aggression in free-ranging male tree lizards (Urosaurus ornatus). Brain Behav Evol 2001; 58:356-361.
35. Lippolis G, Bisazza A, Rogers LJ et al. Lateralization of predator avoidance responses in three species of toads. Laterality 2002; 7:163-183.
36. Robins A, Rogers LJ. Lateralised prey catching responses in the toad (Bufo marinus): Analysis of complex visual stimuli. Anim Behav 2004; 68:567-575.
37. Vallortigara G, Rogers LJ, Bisazza A et al. Complementary right and left hemifield use for predatory and agonistic behavior. Neuroreport 1998; 9:3341-3344.
38. Vallortigara G, Rogers LJ. Survival with an asymmetrical brain: Advantages and disadvantages of cerebral lateralization. Behav Brain Sciences 2005; 28(4):575-633.

CHAPTER 6

Dealing with Objects in Space:
Lateralized Mechanisms of Perception and Cognition in the Domestic Chick (*Gallus gallus*)

Lucia Regolin*

Abstract

The domestic chick constitutes an excellent animal model for the investigation of the lateralization of brain functions possibly underlying a variety of perceptual and cognitive abilities. In particular, lateralized information processing is considered to take place in perception of partly occluded objects, i.e., in the so-called process of amodal completion, and for the knowledge about the existence and the position of objects no longer available to direct perception, particularly when such knowledge is probed in working memory (delayed response task).

Available data indicate that in the domestic chick, the right hemisphere/left eye is in charge of processing amodal completion. Moreover, when engaged in a working memory task, chicks showed right-hemispheric dominance for locating a target on the basis of position-specific cues and bilateral participation of both hemispheres for locating a target on the basis of object specific cues. Interestingly, the results of the experiments with the delayed response task showed that chicks did not exhibit any asymmetry in working memory when position- and object-specific cues were available either separately or together. An asymmetry only appeared when object-specific and position-specific cues were present simultaneously but provided contradictory information, in which case the left-eyed chicks clearly chose the spatial cue, ignoring the object characteristics, whereas the right-eyed chicks chose similarly both the spatial and object cues.

Introduction

Animals move elegantly through their own natural environments, which often consist of very complex sets of objects meaningful to the living creatures. Some of these objects may constitute physical obstacles to the animal's motion while others may instead constitute shelters from predators.

Recent experimental studies confirmed that, in nonhuman animal species, a variety of complex perceptual and cognitive abilities are present, such as: the perception of objects that are not entirely visible because they are partly occluded by other objects, the knowledge about the existence of previously experienced objects once they are no longer directly available to perception, and the knowledge of the position in space of such objects. These skills were shown to be based on lateralized processing of information, i.e., different behavioral patterns are evoked when either one or the other of the cerebral hemispheres is in control.

*Lucia Regolin—Department of General Psychology, University of Padua, Via Venezia 8, 35131 Padova, Italy. Email: lucia.regolin@psico.unipd.it

Behavioral and Morphological Asymmetries in Vertebrates, edited by Yegor B. Malashichev and A. Wallace Deckel. ©2006 Landes Bioscience.

Lateralization of brain functions is well documented in the domestic chick on the basis of behavioral,[1-2] pharmacological,[3-8] and neurobiological[9,10] techniques, which revealed that the two hemispheres differ in fundamental ways in their modes of analysis and storage of perceptual information.

In birds, the fibers of the avian optic nerves cross over nearly completely to the contralateral hemisphere. This anatomical asymmetry allows for the inference of hemispheric dominance during behavior. That is, observations of either binocular or monocular (left or right) eye use during behavior implies control over that behavior by the contralateral cerebral hemisphere.

Procedures that restrict direct sensory input to one or the other hemisphere have proved particularly valuable in the study of asymmetrical cerebral control of behavior. Chicks using their left eye tend to choose between objects to which they are socially attached on the basis of small changes in their appearance. These same changes generally are ignored by chicks using their right eye.[11,12] (Following convention, the term 'hemisphere' will be used here for brevity to stand for the cerebral structures contralateral to the seeing eye; see ref. 2). Chicks using their right nostril (and so predominantly their ipsilateral hemisphere) show a similar pattern in choice based on olfactory changes.[13] Chicks using their left eye also have a striking advantage in topographical orientation based on visual cues.[14] Overall, these findings suggest a special competence of the right hemisphere in spatial analysis and in response to novelty.[2,15,16]

The left hemisphere (right eye in use), in contrast, seems in charge of the selection of features allowing stimuli to be assigned to discrete categories. In fact, a category accommodates a range of different exemplars, sharing a certain proportion of common features, despite a certain degree of variation between such same exemplars in a variety of other properties.[2,15]

In newly hatched chicks (*Gallus gallus*), the role of lateralization was investigated for the control of the process necessary in order to recognize occluded objects. In order to see an object as occluded behind another object, some missing features belonging to the occluded object must be completed and the perceptual process involved has been named "amodal completion": parts that are not physically available to the sense organs, are nevertheless perceived as existing, although hidden.[17] For carrying out this investigation, a group of chicks were imprinted binocularly on a small cardboard square partly occluded by a superimposed bar. At test, in monocular conditions, each chick was presented with a free choice between a complete and an amputated square. Chicks tested during left eye only use chose the complete square, similarly to chicks that were tested during binocular eye use.[18] Right-eyed chicks, in contrast, tended to choose the amputated square. These findings suggested that left eye/right hemisphere system in the chick might process the 'global' analysis of visual scenes, whereas right eye/left hemisphere system may process a "featural" analysis of visual scenes.

A second set of experiments[19] was aimed at examining the lateralized processes possibly involved in the use of working memory. In these experiments an attractive goal-object disappeared completely from chick's sight and working memory had to be used by the chicks in order to find the disappeared goal in the correct spatial location in a series of delayed-response trials. The delayed-response task consists in preventing the subject's response for a given delay. In our experiments chicks could track the goal disappear behind one of two hiding locations but could only perform its search after 30 sec of confinement. For solving the task working memory is necessary, in fact the animal must remember the position of the goal throughout the delay although every chick underwent many trials and in each trial the goal could disappear at random in each of the two spatial locations. (See also Fig. 1 in Chapter 9 by G. Vallortigara.) Either object- or position-specific information could be available to the chick in order to locate the goal that had disappeared behind one of two screens in a test arena. When position-specific information was the only available cue (the two hiding screens were identical to each other) binocular and monocular chicks could easily and equally well locate the goal. Similarly, when only object cues, i.e., the visual characteristics of the screens varied while position-specific cues remained constant (several different screens were alternated for this, more complex, procedure), binocular and monocular chicks could remember the goal.

When both object- and position-specific cues were available to the chick, but were in conflict (i.e., two different screens were used, but their position in space was changed during the delay, when the chick was confined in an opaque cage), only left-eyed and binocular chicks went to the correct position. Right eyed chicks seemed to choose both the correct spatial position and the correct object cue to the same degree. Note that there was no mistaken response in this task (i.e., either the chick went to the correct screen, or to the correct spatial position). Right eyed chicks did not show a preferred strategy, and simply went for both solutions at random. The results suggest that object- and position-specific information is available to both hemispheres in working memory tasks. However, when a conflict between cues arises, the right hemisphere preferentially attends to position-specific cues, whereas the left hemisphere tends to attend to object-specific cues.

Differences in modes of analysis between the cerebral hemispheres provide a unique opportunity to look at the way in which the various aspects of visual representations of objects are organized. These data pave the way for future investigations aimed at clarifying the mechanisms at the base of cognitive and perceptual abilities in the animal—not just avian—brains.

Visual Perception

The Two Hemispheres Differ at Completing Partly Occluded Objects
Human gazing at partly hidden objects, "fill in" for the missing regions of the figure using the perceptual process of 'amodal' completion.[20-23] Amodal completion requires the subject to cognitively generate the nonvisible parts, and this, in turn, depends on detection of certain configurational relationships in visual scenes, such as the alignment of visible parts and similarities in their colors and textures. Perception of object unity in certain partial occlusion displays has been demonstrated in human infants as young as two months of age.[24,25]

Among mammals, evidence suggests that other species such as mice[26] and primates (chimpanzees,[27] rhesus monkeys,[28] and baboons[29]) also show recognition of partly occluded objects. Whether or not birds, as well, utilize amodal completion as they forage and function in their dynamically changing visual environment has not been well studied until recently.

Forkman and colleagues reported that adult hens use amodal completion during conditioning paradigms.[30,31] Also psittacine birds, such as parrots and parakeets,[32,33] mynahs,[34] and magpies[35] use amodal processing to pass, without difficulty, standardized tests of object permanence in which the subject has to respond to partly occluded objects. In contrast, pigeons (*Columba livia*) are poor at amodal processing, although the data is equivocal. Several studies reported that pigeons respond on the basis of local, visible, features[36] and fail to complete[37] or even to perceive[38] the continuation of figures behind an occluder. Other work suggests that perception of object unity occurs in pigeons.[39-41] It seems that pigeons perceive complex stimuli as an assembly of local features and respond to partly occluded objects on the basis of the visual information remaining after fragmentation of the stimulus. However, it has been shown[42] that experimental paradigms using colour slides of conspecifics rather than artificial figures cause pigeons to respond to more generalized perceptual features. Moreover, recent work[43] shows how pigeons are capable of recognizing two-dimensional partly occluded objects if special training is provided.

Investigating the perception of subjective contours, a phenomenon closely linked to amodal completion,[44] Prior and Güntürkün showed[45] that a minority (four out of fourteen) of the pigeons in their study reacted as if they were seeing subjective contours. Pigeons responding to subjective contours seemed to be attending to the "global" pattern of the stimuli, whereas pigeons not responding to subjective contours were attending to single elements.

Overall, these results[43,45] seem to suggest that pigeons can actually perceive and discriminate complex stimuli on the basis of either the local parts or the global configuration, switching, although with some effort, from a "featural" to a "global" style of analysis. The fact that only pigeons attending to the more 'global' aspects of the stimulation responded to subjective contours suggests that such an individual variability in attending 'globally' or 'locally' to visual scenes can account for pigeons' failure in amodal completion tests, which are effective for other species.[37]

In the above mentioned work, the use of conditioning paradigms is likely to have affected the interpretation of the data, as conditioning paradigms may prompt the animal to respond on the base of local, single features of the reinforced configuration. One method of addressing this methodological problem is to replace the conditioning paradigm with that of filial imprinting, a learning process by which global characteristics of objects are acquired by the animal by sheer exposure to them, in the absence of any reinforcement. Regolin and Vallortigara[18] reared chicks singly with a triangle made of red cardboard. At test, on day 3, separate groups of chicks were presented with pairs of stimuli located at the opposite ends of a test-cage. When faced with a choice between a complete and an amputated triangle, chicks clearly preferred to associate with a complete triangle, the stimulus they had been reared with. The choice did not seem to be due to a generic preference for figures with more extended red areas: when the amputated parts of the triangle were dislocated so as to produce a 'scrambled' triangle, chicks still preferred the complete one. When faced with a partly occluded and an amputated triangle (both stimuli presented exactly the same amount of red and black areas), chicks clearly chose the partly occluded triangle. When chicks were reared with a partly occluded shape, the opposite outcome could be observed: chicks exposed to a partly occluded triangle preferred a complete triangle to a fragmented one.

The above findings have been replicated by Lea, Slater, and Ryan[46] using a technique employed by developmental psychologists (see ref. 24). Chicks were imprinted onto three different stimuli, these were: (1) two pieces of rods moving above and below a central occluder, (2) a complete rod, moving in the absence of the occluder, and (3) two pieces of rods also moving in the absence of the occluder. At test, chicks reared with the two pieces of rods and no occluder preferred this stimulus to a complete rod, whereas chicks imprinted with an occluded rod (i.e., only two pieces of rods were present in the actual imprinting stimulus, but the presence of an occluder determined the perceptual completion the two pieces into one whole, although occluded, rod) preferred at test the solid rod over the two pieces of rods.

Overall, these findings favor the idea that birds possess visual perception of partly occluded objects in a fashion similar to humans and primates. Chicks show perception of object unity soon after hatching, but chicks belong to precocial species (i.e., born or hatched highly competent from a sensorimotor perspective, but not only. Newly hatched chicken don't have to learn to walk around and peck for food, as opposite to altricial species). On the other hand, the human infant, who is unable at birth to coordinate motor movements with visual gaze, requires several months for the development and the emergence of complex abilities, readily available, at birth, to the newborn chick.

It is unclear if amodal completion abilities are indeed available to both hemispheres. However, because amodal completion is a basic visual ability necessary for survival, it is reasonable to speculate that both hemispheres would participate in its processing. Corballis and coworkers[47] reported that, following callosotomy, only the human patient's right hemisphere could succeed in amodal completion. Whether this reflects different abilities of the two hemispheres in visual processing or whether it is due to attentional mechanisms remains to be established.

Although there are no direct studies in birds on the effects of callosotomy on amodal processing, there are data on a closely related phenomenon, the perception of subjective contours.[20] It was the Italian psychologist Gaetano Kanizsa who first described an illusion in which subjects clearly reported to see a figure brighter than the surrounding area, usually a white equilateral triangle, wereas, in fact, no triangle was drawn nor was there any difference in brightness between the area occupied by the perceived figure and the actual background. This effect, known as subjective contours, was due to the presence of other, darker, figures, such as discs with a missing sector, which were perceived as whole discs occluded by the brighter figure.

In a study on subjective contours with Kanizsa's triangles and squares Prior and Güntürkün[45] were able to demonstrate that some (but not all) of the pigeons reacted to the test stimuli as if they were seeing subjective contours. This is perhaps not particularly surprising, given that perception of subjective contours in birds has been demonstrated using behavioral methods in the domestic chick[48] and in the barn owl.[49] Moreover, in the barn owl neurons have been found in the visual Wulst whose discharge rate is selectively modulated by subjective contours. It is interesting, however, that only a minority of pigeons responded to subjective contours. As indicated by control tests, pigeons responding to subjective contours were attending to the "global" pattern of stimuli, whereas pigeons not responding to subjective contours were attending to extracted elements of the stimuli. These findings suggest that perception of subjective contours is closely linked to amodal completion. In natural situations in which objects occlude one another, boundaries may vanish and interpolation mechanisms to reconstruct missing contours are sometimes needed. The ability of an organism to attend "globally" versus "locally" to visual scenes, which differs between pigeons and mammals, may explain why pigeons sometimes failed in amodal completion tests which are successfully performed in mammals.[37] Although the pattern emerging in monocular left and right stimulation in the study by Prior and Güntürkün[45] were similar, suggesting that in pigeons that responded to subjective contours both the left and the right hemispheres were capable of 'filling in' processes, it could be that some basic difference between the hemispheres nonetheless exists as a function of pigeons that do, and do not respond to subjective contours. For instance, it could be that dominance of one or other hemisphere favors more 'global' or 'local' strategy of analysis of visual stimuli.

We employed the imprinting paradigm to test for recognition of partly occluded objects in monocularly tested birds. In the first experiment we aimed at investigating the presence of lateralization for amodal completion in chicks by replicating the experiment by Regolin and Vallortigara.[18] This experiment used chicks imprinted in binocular condition who subsequently were tested with only their left or right eye in use. Subjects included 160 female Hybro *Gallus gallus* chicks incubated and hatched in identical and standard conditions. Subjects were housed singly in cages together with an imprinting stimulus pasted onto one of the shorter walls of the cage, at about the chick's head height.

Each chick was exposed, in its home cage, to one single imprinting object. Three different objects were used, hence there were three different experimental groups (each chick was randomly assigned to one of the three experimental groups). The three different imprinting stimuli (represented in Fig. 1) were: (a) a red cardboard square, partly occluded by a black rectangular cardboard bar superimposed on it; (b) a whole red square identical to that used for condition "a" but without any occluder present; (c) an "amputated square" made only by the two visible parts of stimulus "a" separated by a central gap (for more detail on the stimuli employed see ref. 18).

Chicks from each imprinting group were randomly assigned to one of two testing conditions: left eye versus right eye use. Overall, 79 chicks were temporarily patched on their left eye, and 81 chicks were temporarily patched on their right eye. The test apparatus (for detailed description see ref. 11 and Fig. 5 in Chapter 9 by G. Vallortigara) consisted of a runway 45 cm long, 20 cm wide and 30 cm high. The runway was divided, by two fine lines drawn on the floor, into three virtual compartments including one central and two side compartments. At each end of the runway one of the test stimuli could be seen. For all chicks the test consisted of a choice between a whole square identical to imprinting stimulus "b" placed on one end of the test apparatus, and an "amputated" version of the square identical to imprinting stimulus "c" on the other end of the apparatus. Each chick was in turn placed in the central portion of the test corridor and its behavior was thereafter observed for six consecutive minutes. The time (in seconds) spent in each of the three compartments was recorded and computed for each chick as a percentage of time spent by the whole square. Departures from chance level indicated either a preference for the whole square or for the "amputated" square.

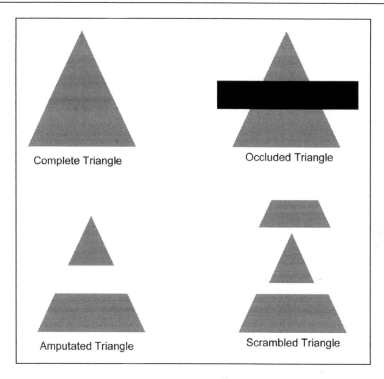

Figure 1. The stimuli employed by Regolin and Vallortigara[18] for the investigation of amodal completion.

Results (Fig. 2) indicated that, on average, chicks imprinted onto the partly occluded square when using their left eye tended to prefer the complete square ($p < 0.01$). Chicks using their right eye, on the other hand, did not show a clear choice, though they seemed to prefer the amputated square ($p > 0.1$). Results for chicks imprinted on the complete square showed a significant preference by all chicks for choosing the complete square over the amputated square ($p < 0.001$). Results for chicks imprinted on the amputated square indicated, overall, a significant preference for the amputated square over the complete square ($p < 0.0001$).

Overall, results show that left-eyed chicks are capable of amodal completion, replicating previous work using binocular chicks.[18,46] Chicks imprinted on the complete square preferred the complete square to the amputated one, while chicks imprinted on the amputated square preferred the amputated square to the complete one. Thus the chicks' choice depends on previous imprinting on a certain configuration, and not on preferences for novelty and/or larger colored area. Chicks imprinted on the partly occluded square and tested with their left eye preferred, however, the complete square to its amputated version. This occurred in spite of the fact that the chicks were exposed to an imprinting stimulus physically identical to the amputated square as for the shape and extension of the red surface. Thus, left-eyed chicks seemed to "complete" partly occluded objects as binocular chicks did.[18,46] Right-eyed chicks, in contrast, did not show any clear evidence of amodal completion. Rather, they exhibited a trend to choose the amputated square, as if they were imprinted on the amputated rather than onto the occluded stimulus.

These results suggest that, in the domestic chick, the right hemisphere/left eye either is in charge of processing amodal completion or, alternatively, is involved with the chick's ability to focus its attention onto the global configuration of the visual percept. Conversely, the data suggest that the right-eyed chicks pay greater attention to the local features of the stimulus.

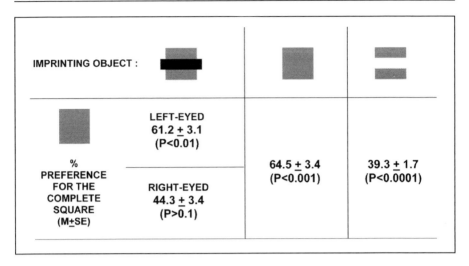

Figure 2. Results of the experiment by Regolin et al[17] on the presence of lateralized processes for amodal completion; Mean ± SE, one-sample *t*-test to compare the data with chance level (50%).

Results from this first experiment on the biological bases of amodal completion were confirmed in a second one, for which we used a different type of stimulus to control for any effect due to stimulus symmetry and orientation, after occlusion. A cross was employed instead of a square as this shape remains symmetrical along both the horizontal and vertical axis after occlusion by a round black patch. Chicks imprinted onto the amputated cross tended to prefer the amputated cross (the choice for the amputated cross was significant if considering the last two minutes of the test; Mean ± SE one-sample *t*-test: 41.54 ± 4.41, p < 0.05), this in spite of the larger amount of red area present in the comparison stimulus (the whole cross). On the other hand, binocular and left eyed chicks imprinted onto the patched cross preferred the amodally completed cross (61.78 ± 3.96, p < 0.01; the pattern of choice became even clearer in the last two minutes of the test: 75.47 ± 6.09, p < 0.001). This preference did not extend to right-eyed chicks, which did not exhibit any clear preference, although overall tended to approach the amputated stimulus (45.61 ± 6.45, p > 0.1).

In summary, the above experiment generally replicates our initial findings but data became clear only during the last minutes of the experiment. Perhaps the process of completion for the stimuli used in the second experiment is more difficult due to the relatively larger occluded area in the cross as compared to the square. The occluded area, in the case of the patched cross, corresponds to 43% of its surface, i.e., about 10% more than in the case of the occluded square, where only 33% of its surface is covered by the bar. It is also possible that a part of the difference in choice is masked by differences in the time spent in the middle compartment (in both testing conditions). Right-eyed chicks showed longer times spent in the middle compartment that was crucially involved in amodal completion, possibly reflecting difficulty in choosing between the two stimuli.

From Visual Perception to Spatial Cognition

Lateral Bias at Detouring Obstacles

Sometimes objects are totally concealed behind other objects. Yet, even when an object disappears completely from sight because of an obstacle, we can attribute continued existence to the disappearing object. A number of studies have investigated these issues of animal cognition in apes and monkeys,[50,51] cats,[52-56] and dogs.[53,57] These studies demonstrated that the capacity of at least some nonhuman mammals (e.g., chimpanzees [*Pan troglodytes*], gorillas [*Gorilla gorilla*] and dogs [*Canis familiaris*]) for solving 'object permanence' problems is comparable to that of humans (see refs. 58,59 for reviews).

Detour behavior, i.e., the development of itineraries that allow for recognizing obstacles between subject and goal, is of particular interest with respect to the issues of object representation and object permanence. Köhler[60] first introduced this problem as a test of insight learning. Detour performance, investigated extensively in human infants,[61] suggests the mental representation of at least some of the characteristics of the object that has disappeared. Köhler[60] reported that chicken had difficulty in detour problems. This has been traditionally interpreted as indicating that chicken have a poor ability to form cognitive maps. According to some researchers[62-64] chicks learn the correct route to a goal after repeated trials, but have difficulties in solving the problem on the first attempt. More recent work,[65] though, shows that chick's difficulties in detour problems can be ascribed to the motivational overtones and perceptual ambiguities of the experimental situation. In fact, the task difficulty can be dramatically changed simply by modifying the characteristics of the obstacle or of the goal, i.e., the less visible the goal, the easier the task. Moreover, apart from the attractiveness of the goal itself, there seem to be barriers that are perceptually less of an obstacle than others. For example, they may not be true obstacles in a natural environment for that species (e.g., vertical barriers) or, alternatively, they are somewhat special and not normally encountered in a natural environment (e.g., transparent barriers). Social signals emitted by the goal also appear to be a crucial factor.

Chicks as young as two days after hatching do possess the cognitive abilities required by a detour behavior task. Some sort of representation of the goal object and its spatial location seem to be necessary to account for the chicks' performance. Do chicks really have the ability to represent the goal and its spatial location in the absence of locally orienting cues? The issue could be addressed by looking at the chick's detour behavior during its very first attempt to rejoin the goal after its disappearance (to rule out the effect of previous experience, which could affect chick's behavior in subsequent trials) and in the complete absence of sensory cues (to rule out the effect of such cues in orienting the chick towards the goal). If, in such conditions, chicks move randomly in the environment when the goal is no longer available to direct perception, then no straightforward conclusion can be drawn because chicks may represent the object but lack any ability to discover its position. If, on the other hand, chicks move nonrandomly and show an ability to orient towards the disappeared goal, then some sort of memory for mental representation of the goal in a certain spatial location can be inferred. This, of course, leaves open the issue of what the nature of the mechanism to localize the goal would be. Our data[66] clearly show that chicks are able to turn correctly towards a goal in the absence of any locally orienting cues. They appear to remember the location (and thus the presence) of a social partner even when this was no longer available to direct perception.

Another important issue concerns the type of goal-object used in these studies, such as food vs. social partners. It may be that in a natural environment different goals elicit different searching behavior strategies, and that these strategies affect the probability of a chick solving detour problems in the laboratory. Alternatively, different goals may be associated with the triggering of different emotional and/or motivational responses, some of which could interfere with the execution of the task and mask the true cognitive abilities of the animal. Using a mealworm as a goal, Regolin and coworkers[67] replicated Etienne's work.[62] That is, even with longer testing times and a very large sample, chicks appeared unable to make use of the directional cue provided by the prey movement. Using an artificial social partner as a goal, however, the majority of the chicks were able to choose the correct screen. Clearly, the use of such different goals results in relevant differences in motivational and emotional variables.

Informal observations during our detour tests suggested that chicks could be looking at the goal using one eye preferentially. Specifically, we found that the direction of a detour around a barrier strictly reflected contralateral eye use during the detour. For example, a chick which detoured on the right preferentially used the left eye, while a chick which detoured on the left preferentially used the right eye. We also studied chicks wearing eye patches to check whether or not the two eye systems differed in their detour learning abilities. In the chick, visual lateralization can be tested even when both eyes are unobstructed by simply recording preferential eye use in the viewing of various stimuli. Past work has shown that lateral fixation with one eye activates the contralateral cerebral hemisphere.[68,69] Thus, for instance, McKenzie and coworkers[70] found that, on day 1 post hatch, females (but not males) fixate with the right eye rather than the left eye when naive and deciding whether to approach an imprinting object. Asymmetries were also found when groups of two-, three-, and four-day old chicks were tested binocularly and monocularly using a familiar imprinted red ball located behind a barrier.[71] A small but consistent effect at all ages studied and in both sexes was that right-eyed chicks took less time to detour the barrier than left-eyed chicks. Additionally, binocular chicks showed a bias to detour the barrier on the left side, consistent with preferential right eye use. Responses of binocular chicks were not random, but more similar to those of chicks that used the right eye. It seems very unlikely that these asymmetries were due to a motor bias, because the direction in which the chick turned could be reversed by simply changing the visual characteristics of the red imprinting ball (e.g., yellow, blue, half yellow-half red). There was a shift to left eye use with some of the novel colors (see also refs. 13,16). The shift in eye use depended on an estimation of the degree of novelty of the unfamiliar ball and, interestingly, the judgment seemed to differ between males and females. In females there was a shift from right to left eye use with an increasing degree of novelty. Results for males were puzzling, in that they used the right eye when presented with a blue ball that was very novel to them. An entirely speculative but interesting possibility is that chicks use the right eye (i.e., the left hemisphere) in order to minimize fear (fear responses are known to be under the control of the right hemisphere) due to large transformations in the imprinting objects: females would do it in order "to ignore" the pattern change and males in order 'to ignore' the change occurring when the object is substituted with an identical one but blue in color (such color is probably fearful for the chicks, see refs. 72,73). Previous work had shown that in monocularly tested chicks the left eye is mainly involved in social discrimination between conspecifics.[11,12] We wondered, therefore, whether a preference for using the left eye in estimating novelty could result in a bias to detour the barrier on the right side when chicks were faced with unfamiliar conspecifics.

On the Cognitive Side

The Two Hemispheres Differ in a Working Memory Task

A very limited amount of information is available for the role of the two hemispheres of the avian brain during working memory tasks. Clayton and Krebs[74] tested the memory of food-storing and nonfood-storing birds for feeders that had a trial-unique location in an experimental room as well as a trial-unique color pattern. When, after a short retention interval, birds were given dissociation tests in which the correct feeder changed its position and a different feeder was placed at the original location, all birds preferentially searched using position-based cues when tested with only their left eye and using feeder-specific cues when tested with only their right eye. These results seem to correspond quite closely with evidence obtained in reference memory tasks (see above). However, Prior and Güntürkün[75] trained pigeons to search for food in a maze in a spatial working memory task and in an object-specific working memory task. They found that an object-specific working memory task mainly involved the left hemisphere while a spatial working memory task required the use of both hemispheres.

We developed a technique[76] for studying working memory in chicks using a delayed response task. An attractive goal (an imprinted object) was hidden behind one of two different opaque screens and the chicks were allowed to search for the goal after different time delays. The chicks were exceptionally good at this task, retaining the location of the goal up to intervals of 60 sec.

We have used the same experimental procedure to investigate whether lateralization only occurs in working-memory tasks when a conflict arises between position-specific and object-specific information.[19] (See also Fig. 1 in chapter by Vallortigara) The subjects were 180 *Gallus gallus* chicks that had been reared singly with a small red plastic ball as an imprinting object. The test apparatus consisted of a circular arena (95 cm in diameter) within which was positioned a small box where the chick was confined for a given delay period during the test phase. An opaque partition was used during the test, in order to prevent the chick from seeing the arena during the delay time. Two opaque cardboard screens were positioned in the centre of the arena, symmetrically with respect to the confining box, 20 cm apart from each other and 35 cm away from the closest side of the confining box. The screens varied as a function of color and patterning. Testing took place on day 4 post hatch. The chick was confined in the small cage, behind the clear glass sheet, from where it could see and track the ball disappearing behind one of the two screens. After the disappearance of the ball, an opaque partition was located in front of the glass sheet, in such a way of preventing the chick from seeing the two screens for a time delay of 30 sec. After the delay, the chick was allowed to search for the ball. The first screen approached and circled around by the chick was recorded as either correct or incorrect. Each chick underwent 16 consecutive trials. There were several experimental conditions.

In the first condition ("position cue"; N = 30) only the positional cue (left vs right) was available to the chick in order to identify the correct screen. Half of the subjects saw two blue screens with a yellow 'X', while the other half saw two white screens with a red pattern.

In the second condition ("object cue"; N = 30) only nonspatial cues, consisting of several screens of different colour and patterns, were available to the chick. In each trial the chick saw the ball disappearing behind one single, centrally located screen, much like what happened during the training phase. During the delay, while the chick's sight was blocked by the opaque partition, a second screen of a different colour was introduced in the arena, and the two screens were positioned in the standard testing fashion described above. The chick faced now two screens, and was required to identify the one behind which the ball had disappeared. This task required the chick to rely only on the object cues. From trial to trial the correct screen alternated from left to right with respect to the chick's position. A total of six different screens were used including four new screens in addition to the two described for the previous condition. The same screen never appeared in two consecutive trials. All screens appeared several times on both sides of the chick, and sometimes were, or were not, the correct target.

The third condition ("position plus object cues"; N = 30) required that both position and colour could be used to identify the correct screen. The same two screens used in condition 1, which were different in colour and pattern, were employed for condition 3. One screen was blue with a yellow 'X', and the other—white with a red pattern.

In the fourth condition ("contrasting position and object cues"; N = 42) the two screens were similar to those used in condition 3. However, in condition 4 the screens were swapped during the delay period. This required the chick to choose between a screen that was in the correct position but had the wrong colour and patterning, and a screen that was in the wrong position but had the correct colour and patterning. In condition 4, by convention, the choice of the correct position was considered correct.

For each of the four conditions, separate groups of chicks were tested in binocular and monocular conditions. Monocular testing was carried out by means of temporary eye patches. The chick was considered to have made a choice when it circled around one of the screens. The percentage of correct choices was computed for each chick as the (number of correct choices/ total number of choices expressed by the chick) × 100.

Results are shown in Figure 3. Overall, in the first three conditions (condition 1, condition 2 and condition 3) chicks were always able to solve the task by successfully identifying the correct screen, with no differences between binocular, right-, and left-eyed chicks.

CONDITION 1 (spatial cue)	CONDITION 2 (color cue)	CONDITION 3 (spatial+color)	CONDITION 4 (spatial against color)
All chicks could equally successfully locate the goal	All chicks could equally successfully locate the goal	All chicks could equally successfully locate the goal	Main effect of Eye $P = 0.0001$ Binocular : 69.593 ± 2.531 $P < 0.0001$
71.688 ± 2.236 $P < 0.0001$	60.907 ± 1.818 $P < 0.0001$	71.177 ± 1.248 $P < 0.0001$	Left-eyed : 73.054 ± 2.904 $P < 0.0001$ Right-eyed : 51.133 ± 3.398 $P = 0.744$

Figure 3. Results of the experiment by Regolin et al[19] on the presence of lateralized processes for a task involving working memory (delayed response); Mean ± SE, one-sample *t*-test to compare the data with chance level (50%).

In particular, when two identical screens were used (condition 1) chicks could successfully locate the goal (p < 0.0001) by identifying the correct screen on the basis of spatial information alone. Also when the color and texture of the correct screen were the only relevant information in order to identify it (condition 2), chicks could successfully locate the goal (p < 0.0001). Finally, when chicks could use both color and position cues (condition 3), all chicks could successfully locate the correct screen (p < 0.0001).

Condition 2 turned out to be more difficult for the chicks than condition 1 (p = 0.0004) or condition 3 (p < 0.0001). Moreover, condition 3 did not seem to be easier to solve than condition 1, thus adding an objectual cue (e.g., color) to the spatial cue already present in condition 1 did not seem to improve performance. This is consistent with previous evidence of a primacy of position cues in visual discrimination learning in chicks.[77-79]

In condition 4, two different screens were used and the screen with the "correct" object-characteristics (the one, behind which the ball had been hidden) was changed with the incorrect screen, and hence the correct screen was located in the wrong spatial position at the end of the delay. The analysis revealed a laterality effect. The performance of binocular and left-eye chicks did not differ (p = 0.424). Conversely, performance was significantly worse when right eye chicks were compared to both binocular (p = 0.0001) and left-eyed (p = 0.0001) groups.

In condition 4, at test, both binocular (p < 0.0001) and left-eyed chicks (p < 0.0001) approached the correct spatial position (this means, of course, the screen with "incorrect" visual characteristics) whereas right-eyed chicks chose at random (p < 0.744) between correct spatial position and correct object-cue (i.e., color of the correct screen).

Similar results for reference memory were recently reported.[80,81] Chicken were trained to find food by ground scratching in a closed uniform arena that had a red stick placed in it. After binocular training in this arena, the red stick landmark was displaced in a corner, so that object-specific cues (the landmark) and position-specific cues (the central position) provided contradictory information. A striking asymmetry resulted, binocular and left-eyed chicks searched at the center (ignoring the landmark), whereas right-eyed chicks searched at the corner (ignoring purely spatial information). There were, however, no differences in the ability to retrieve the correct spatial location between right- and left-eyed chicks after binocular training. If, after binocular training, the central landmark was removed, both eye-systems were able to search in the correct spatial (central) position. Similarly, in the delayed response task, chicks tested with two identical screens or with two screens of different colors that maintained fixed spatial position during the various trials, showed no asymmetry when tested with only the right or left eye. It seems that the prevalent use of spatial cues by the left eye and of object cues by the right eye only emerges when these cues are simultaneously available and provide conflicting information. This could be explained by supposing that, even when only an eye is in use in the chick, information stored in the ipsilateral hemisphere could be nonetheless assessed and employed to control behavior (for instance, to retrieve spatial information stored into the right hemisphere even when vision is confined to the right eye-left hemisphere only).

When position- and object-specific cues are simultaneously available, however, the hemisphere which is directly stimulated, controls behavior by relying on its 'preferred' cues, thus giving rise to choice of position-specific cues for the left eye and object-specific cues for the right eye. This is consistent with recent neuroanatomical evidence. The primary visual projections generally ascend contralaterally in the chick's brain. However, minor ipsilateral as well as major contralateral projections are both present in the thalamofugal as well as in the tectofugal pathways.[82-84] Thus, behavioral paradigms that use monocular vision in chicks are difficult to interpret from the perspective of lateralized control of behavior.

General Discussion and Conclusion

The results of the first set of experiments suggest that the right hemisphere/left eye is mainly responsible for the process of the "amodal" completion of partly occluded objects. In order to amodally complete an object, the brain must connect and fill in the parts that are missing in a visual scene, a task in which the right hemisphere is known to be very good at.[11,85,86] Besides this, chicks that use only their left eye behave quite similarly to binocular chicks. Thus, the right hemisphere may be more specialized at detecting the global structure of visual objects, whereas the left hemisphere may be better at detecting local features. Moreover, the right hemisphere seems to be the hemisphere in charge of control of behavior in these tasks, as evinced by the fact that binocular chicks behaved similarly to left-eyed chicks.

Which hemisphere controls overt behavior can obviously change depending on several variables, such as the nature of the task and the motivational/emotional overtones associated with it. There may be species differences as well. This is particularly intriguing with regards to the data collected in the pigeon, a species with a different organization of the visual pathway as compared to the chick.[87] Pigeons, as is well documented (review in ref. 88), show left hemisphere dominance during visual discrimination tasks in binocular conditions. This may well predispose them towards a featural, rather than a global, style of analyses of visual scenes, making amodal completion difficult to observe.

Of course, factors aside from amodal completion, such as stimulus size and contour continuity[18,46,89] may account for the differences between the groups. It is unavoidable that several physical parameters change in the stimuli from exposure to test. In the experiments described chicks simply go to the imprinted stimulus, irrespective of its having a larger or a smaller red area. The crucial condition is that of amodal completion: here no physical similarity with the training stimulus can be predicted a priori. However, the behavior of the chicks can tell us what stimulus they judge to resemble more to the training (exposure) stimulus. Binocular and left-eyed, but not right-eyed, chicks behave in a way that is consistent with perceptual completion.

Another important issue concerns the type of hemispheric differences with regards to "global" and "local" processing. It is unlikely that the left -eye/right hemisphere is "binding together" all visual stimuli. Rather, the right hemisphere is likely utilizing amodal completion for only those objects that are occluded. We believe that hemispheric differences (as well as the possibly associated species differences) are mostly a matter of degree rather than of kind. Amodal completion is such a crucial mechanism that it is likely to be available to the animal at times. In the natural condition, when birds can use freely both hemifields, the two strategies of holistic and analytic visual analysis should reciprocally support each other rather than compete. Thus, we are inclined to think that hemispheric differences can modulate, probably by attentional mechanisms, the type of analysis to be carried out on visual stimuli. Evidence suggests that birds can bring into action the hemisphere most appropriate to particular conditions and to particular stimuli by using lateral fixation with the contralateral eye.[69,90] Such a mechanism may appear to be very unusual to us, because we are, as mammals, accustomed of using obligatory conjugate eye movements to fixate binocularly any stimulus of interest. However, lateralized mechanisms similar to those available to birds have been described in the human neuropsychology literature, for instance in the form of lateralized direction of gaze or voluntary eye movements to the left or to the right associated with the type of hemispheric strategies to bring into play (for example, see ref. 91).

Interestingly, even in humans the right hemisphere seems to play a more important part in amodal completion. In a case study on two split-brain patients, Corballis et al[92] suggested that amodal completion seems to reflect a high-level lateralized process located in the right hemisphere. It remains to be established whether this reflects different abilities of each hemisphere in early visual processing or, rather, in attentional mechanisms as we have proposed here for avian lateralization.

The results of the experiments with the delayed response task showed that chicks did not exhibit any asymmetry in working memory when position- and object-specific cues were available in isolation, or when there was addition of both of these cues. An asymmetry only appeared when object-specific and position-specific cues provided contradictory information, in which case left-eyed chicks clearly chose the position, ignoring the characteristics of the screen, whereas right-eyed chicks chose at random. Having said this, a tendency in females to prefer object-characteristics was found. Previous work with a sample of females only revealed a slight effect,[93] suggesting a mild preference for object-characteristics at least in one sex. Thus, it seems that working memory tasks reveal a somewhat different pattern of hemispheric specialization than reference memory tasks. In reference memory tasks there is clear evidence for a right-hemisphere dominance for spatial cues and a left-hemisphere dominance for object-specific cues such as colour.[86,94] This is apparent even in the absence of any contrast between different sources of information (see ref. 95 for a review). However, one exception is given by some recent work on lateralization of the so-called "geometric module". Vallortigara et al[94] trained chicks binocularly in an environment with a distinctive geometry (a rectangular cage) with panels at the corners providing nongeometric cues. Between trials chicks were passively disoriented to disable dead reckoning. When tested after removal of the panels, left-eyed chicks, but not right-eyed chicks, reoriented using the residual information provided by the geometry of the cage. When tested after removal of geometric information (e.g., in a squareshaped cage), both right- and left-eyed chicks reoriented using the residual nongeometrical information provided by the panels. When trained binocularly with only geometric information, at test left-eyed chicks reoriented better than right-eyed chicks. However, when geometric and nongeometric cues provided contradictory information, left-eyed chicks showed more reliance on geometric cues, whereas right-eyed chicks showed more reliance on nongeometric cues. The results suggest separate mechanisms for dealing with spatial reorientation problems, with the right hemisphere taking charge of large-scale geometry of the environment and with both hemispheres taking charge of local, nongeometric cues when available in isolation. These findings also suggest a predominance of the left hemisphere activation when competition between geometric and nongeometric information occurs. Similarly, the right hemisphere does not reveal any dominance when the cues are available in isolation, but is dominant when competition between object- and position-specific cues occurs.

It is likely that, under short intervals of retention, information concerning the "what" and "where" is retained in both hemispheres. Lateralization, however, seems to arise even in working memory when a decision about the use of available information should be undertaken. In these cases, the right hemisphere, which has an intrinsic ability to attend to spatial cues, controls behavior as demonstrated by the fact that binocular chicks behave as left-eyed chicks. From a comparative perspective, it is unclear whether it is the nature of the task or species-differences that accounts for why there is dominance of one hemisphere for a particular cue and participation of both hemispheres for others. For instance, in working memory tasks pigeons showed left-hemispheric dominance for object-specific cues and participation of both hemispheres for position-specific cues.[75] In contrast, chicks showed right-hemispheric dominance for position-specific cues and (largely) bilateral participation of both hemispheres for object specific cues (condition "contrasting position and object cues"). In a similar condition, tits (Paridae, Passeriformes) showed complementarities of function, with the right hemisphere attending to position cues and the left hemisphere attending to object cues.[74] In the chick and in the pigeon there is evidence[5,96,97] that in the natural condition the embryo is oriented in the egg so that the right eye is exposed to light (and the left eye is occluded) and this puts the left hemisphere in charge of certain visually-guided patterns of behavior.[98] The chicks used in the present experiments came from a commercial hatchery, in which no light exposure to the eggs was provided. Note, also, that chicks are a precocial species, whereas parids and pigeons are altricial species, and this may also contribute to behavioral differences. Nonetheless, it is worth noting that the basic pattern of hemispheric specialization remains the same in the various species, for example, in no case a dominance of the right hemisphere for object-specific cues has been reported. Therefore it is likely that each hemisphere retains its basic specialization, though different species, tested in various settings, may show some variability in inter-hemispheric dominance.

References

1. Andrew RJ. Lateralization of emotional and cognitive function in higher vertebrates, with special reference to the domestic chick. In: Ewert JP, Capranica RR, Ingle DJ, eds. Advances in Vertebrate Neuroethology. New York: Plenum Press, 1983:477-505.
2. Andrew RJ. The chick in experiment: Techniques and tests. General. In: Andrew RJ, ed. Neural and Behavioral Plasticity. Oxford: Oxford University Press, 1991:6-11.
3. Rogers LJ. Light experience and asymmetry of brain function in chickens. Nature 1982; 297:223-225.
4. Rogers LJ. Laterality in animals. Int J Comp Psychol 1989; 3:6-25.
5. Rogers LJ. Development of lateralization. In: Andrew RJ, ed. Neural and Behavioral Plasticity: The use of the Chick as a Model. Oxford: Oxford University Press, 1991:507-535.
6. Rogers LJ. The development of brain and behavior in the chicken. Wallingford: CAB International, 1995:1-273.
7. Rogers LJ. Evolution and development of brain asymmetry, and its relevance to language, tool use and consciousness. Int J Comp Psychol 1995; 8:1-15.
8. Rogers LJ, Anson JM. Lateralization of function in the chicken forebrain. Pharmacol Biochem Behav 1979; 10:679-686.
9. Horn G. Neural basis of recognition memory investigate through an analysis of imprinting. Philos Trans R Soc Lond B Biol Sci 1990; 329:133-142.
10. Rose SPR. The Making of Memory: From Molecules to Mind. London: Bantam Press, 1992.
11. Vallortigara G, Andrew RJ. Lateralization of response by chicks to change in a model partner. Anim Behav 1991; 41:187-194.
12. Vallortigara G. Right hemisphere advantage for social recognition in the chick. Neuropsychologia 1992; 30:761-768.
13. Vallortigara G, Andrew RJ. Olfactory lateralization in the chick. Neuropsychologia 1994; 32:417-423.
14. Rashid N, Andrew RJ. Right hemisphere advantages for topographical orientation in the domestic chick. Neuropsychologia 1989; 27:937-948.
15. Andrew RJ, Mench J, Rainey C. Right-left asymmetry of response to visual stimuli in the domestic chick. In: Ingle DJ, Goodale MA, Mansfield RJ, eds. Analysis of Visual Behavior. Cambridge: MIT Press, 1982:225-236.
16. Vallortigara G, Andrew RJ. Differential involvement of right and left hemisphere in individual recognition in the domestic chick. Behav Processes 1994; 33:41-58.

17. Regolin L, Marconato F, Vallortigara G. Hemispheric differences in the recognition of partly oc-
    cluded objects by newly-hatched domestic chicks (Gallus gallus). Anim Cogn 2004; 7(3):162-170.
18. Regolin L, Vallortigara G. Perception of partly occluded objects by young chicks. Percept Psychophys
    1995; 57:971-976.
19. Regolin L, Garzotto B, Rugani R et al. Working memory in the chick: Parallel and lateralized
    mechanisms for encoding of object- and position-specific information. Behav Brain Res 2005;
    157(1):1-9.
20. Kanizsa G. Organization in Vision. New York: Praeger, 1979.
21. Michotte A. The Perception of Causality. New York: Basic Books, 1963.
22. Michotte A, Thinés G, Crabbé G. Les complements amodaux des structures perceptives. Louvain,
    Belgium: Publications Universitaires de Louvain, 1964.
23. Grossberg S, Mingolla E. Neural dynamics of form perception: Boundary completion, illusory
    figures, and neon colour spreading. Psychol Rev 1985; 92:173-211.
24. Kellman PJ, Spelke ES. Perception of partly occluded objects in infancy. Cognit Psychol 1983;
    15:483-524.
25. Johnson SP, Aslin RN. Perception of object unity in 2-month-old infants. Dev Psychol 1995;
    31:739-745.
26. Kanizsa G, Renzi P, Conte S et al. Amodal completion in mouse vision. Perception 1993;
    22:713-721.
27. Sato A, Kanazawa S, Fujita K. Perception of objects unity in chimpanzees (Pan troglodytes). Jpn
    Psychol Res 1997; 39:191-199.
28. Osada Y, Schiller PH. Can monkeys see objects under condition of transparency and occlusion?
    Invest Ophth Vis Sci 1994; 35:1664.
29. Deruelle C, Barbet I, Dépy D et al. Perception of partly occluded figures by baboons (Papio
    papio). Perception 2000; 39:1483-1497.
30. Forkman B. Hen use occlusion to judge depth in a two-dimensional picture. Perception 1998;
    27:861-867.
31. Forkman B, Vallortigara G. Minimization of modal contours: An essential cross-species strategy in
    disambiguating relative depth. Anim Cogn 1999; 2:181-185.
32. Funk MS. Development of object permanence in the New Zealand parakeet (Cyanoramphus
    auriceps). Anim Learn Behav 1996; 24:375-383.
33. Pepperberg IM, Funk MS. Object permanence in four species of psittacine birds: An African Grey
    parrot (Psittacus erithacus), an Illiger mini macaw (Ara maracana), a parakeet (Melopsittacus
    undulatus), and a cockatiel (Nymphicus hollandicus). Anim Learn Behav 1990; 18:97-108.
34. Plowright CMS, Reid S, Kilian T. Finding hidden food: Behavior on visible displacement tasks by
    mynahs (Gracula religiosa) and pigeons (Columba livia). J Comp Psychol 1998; 112:13-25.
35. Pollok B, Prior H, Güntürkün O. Development of object permanence in food-storing magpies
    (Pica pica). J Comp Psychol 2000; 114:148-157.
36. Cerella J. The pigeon's analysis of pictures. Pattern Recogn 1980; 12:1-6.
37. Sekuler AB, Lee JAJ, Shettleworth SJ. Pigeons do not complete partly occluded figures. Perception
    1996; 25:1109-1120.
38. Fujita K. Perceptual completion in rhesus monkey (Macaca mulatta) and pigeons (Columba livia).
    Percept Psychophys 2001; 63:115-125.
39. Towe AL. A study of figural equivalence in the pigeon. J Comp Physiol Psychol 1954; 47:283-287.
40. Hamme LJ van, Wasserman EA, Biederman I. Discrimination of contour-deleted images by pi-
    geons. J Exp Psychol Anim Behav Process 1992; 18:387-399.
41. White KG, Alsop B, Williams L. Prototype identification and categorization of incomplete figures
    by pigeons. Behav Processes 1993; 30:253-258.
42. Watanabe S, Ito Y. Discrimination of individuals in pigeons. Bird Behav 1991; 9:20-29.
43. Di Pietro NT, Wasserman EA, Young ME. Effects of occlusion on pigeon's visual object recogni-
    tion. Perception 2002; 31:1299-1312.
44. Shipley TF, Kellman PJ. Strength of visual interpolation depends on the ratio of physically speci-
    fied to total edge length. Percept Psychoph 1992; 52:97-106.
45. Prior H, Güntürkün O. Patterns of visual lateralization in pigeons: Seeing what is there and be-
    yond. Perception 1999; 28(Suppl):22.
46. Lea SEG, Slater AM, Ryan CME. Perception of object unity in chicks: A comparison with human
    infant. Infant Behav Dev 1996; 19:501-504.
47. Corballis PM, Fendrich R, Shapley R et al. A dissociation between illusory contour perception
    and amodal boundary completion following callosotomy. J Cogn Neurosci 1998; 10(Suppl):18.
48. Zanforlin M. Visual perception of complex forms (anomalous surfaces) in chicks. Italian Journal
    of Psychology 1981; 1:1-16.

49. Nieder A, Wagner H. Perception and neuronal coding of subjective contours in the owl. Nat Neurosci 1999; 2:660-663.
50. Mathieu M, Bouchard MA, Granger L et al. Piagetian object-permanence in Cebus capucinus, Lagothrica flavicauda and Pan troglodytes. Anim Behav 1976; 24:585-588.
51. Natale F, Antinucci F, Spinozzi G et al. Stage 6 object concept in non human primates cognition: A comparison between gorilla (Gorilla gorilla gorilla) and Japanese macaque (Macaca fuscata). J Comp Psychol 1986; 100:335-339.
52. Gruber HE, Girgus JS, Banuazizi A. The development of object permanence in the cat. Dev Psychol 1971; 4:9-15.
53. Triana E, Pasnak R. Object permanence in cats and dogs. Anim Learn Behav 1981; 9:135-139.
54. Doré FY. Object permanence in adult cats (Felix catus). J Comp Psychol 1986; 100:340-347.
55. Doré FY. Search behavior in cats (Felix catus) in an invisible displacement test: Cognition and experience. Can J Psychol 1990; 44:359-370.
56. Dumas C, Doré FY. Cognitive development in kittens (Felix catus): A cross-sectional study of object permanence. J Comp Psychol 1989; 103:191-200.
57. Gagnon S, Doré FY. Search behavior in various breeds of adult dogs (Canis familiaris): Object permanence and olfactory cues. J Comp Psychol 1992; 106:58-68.
58. Etienne SA. Age variability shown by domestic chicks in selected spatial tasks. Behavior 1974; 50:52-76.
59. Dumas C. Object Permanence in cats (Felix catus): An ecological approach to the study of invisible displacements. J Comp Psychol 1992; 106:404-410.
60. Köhler W. The Mentality of Apes. New York: Harcourt Brace, 1925.
61. Piaget J. Origin of intelligence in the Child. London: Routledge and Kegan Paul, 1953.
62. Etienne SA. Searching behaviour towards a disappearing prey in the domestic chick as affected by preliminary experience. Anim Behav 1973; 21:749-761.
63. Scholes NW. Detour learning and development in the domestic chick. J Comp Physiol Psychol 1965; 60:114-116.
64. Scholes NW, Wheaton LG. Critical period for detour learning in developing chicks. Life Sci 1966; 5:1859-1865.
65. Regolin L, Vallortigara G, Zanforlin M. Detour behavior in the chick: A review and reinterpretation. Atti e Memorie dell'Accademia Patavina di Scienze, Lettere ed Arti. Classe di Scienze Matematiche e Naturali 1994; 105:105-126.
66. Regolin L, Vallortigara G, Zanforlin M. Object and spatial representations in detour problems by chicks. Anim Behav 1995; 49:195-199.
67. Regolin L, Vallortigara G, Zanforlin M. Detour behavior in the domestic chick: Searching for a disappearing prey or a disappearing social partner. Anim Behav 1995; 50:203-211.
68. Andrew RJ, Dharmaretnam M. Lateralization and strategies of viewing in the domestic chick. In: Zeigler HP, Bishof HJ, eds. Vision, Brain, and Behavior in Birds. Cambridge: MIT Press, 1993:319-332.
69. Dharmaretnam M, Andrew RJ. Age- and stimulus-specific effects on the use of right and left eyes by the domestic chick. Anim Behav 1994; 48:1395-1406.
70. McKenzie R, Andrew RJ, Jones RB. Lateralization in chicks and hens: New evidence for control of response by the right eye system. Neuropsychologia 1998; 36:51-58.
71. Vallortigara G, Regolin L, Pagni P. Detour behavior, imprinting, and visual lateralization in the domestic chick. Brain Res Cogn Brain Res 1999; 7:307-320.
72. Andrew RJ, Brennan A. The lateralization of fear behavior in the male domestic chick: A developmental study. Anim Behav 1983; 31:1166-1176.
73. Clifton PG, Andrew RJ. The role of stimulus size and colour in the elicitation of testosterone-facilitated aggressive and sexual responses in the domestic chick. Anim Behav 1983; 31:878-886.
74. Clayton NS, Krebs JR. Memory for spatial and object-specific cues in food storing and nonstoring birds. J Comp Physiol A 1994; 174:371-379.
75. Prior H, Güntürkün O. Parallel working memory for spatial location and food-related object cues in foraging pigeons: Binocular and lateralized monocular performance. Learn Mem 2001; 8:44-51.
76. Vallortigara G, Regolin L, Rigoni M et al. Delayed search for a concealed imprinted object in the domestic chick. Anim Cogn 1998; 1:17-24.
77. Zanforlin M, Vallortigara G. Form preferences and stimulus generalization in domestic chicks. Boll Zool 1985; 52:231-238.
78. Vallortigara G, Zanforlin M. Position learning in chicks. Behav Processes 1986; 12:23-32.
79. Vallortigara G, Zanforlin M. Place and object learning in chicks (Gallus gallus domesticus). J Comp Psychol 1989; 103:201-209.

80. Tommasi L, Vallortigara G. Lateralization of spatial memory tasks in the domestic chick (Gallus gallus). Exp Brain Res 1997; 117:S43.
81. Vallortigara G. Comparative neuropsychology of the dual brain: A stroll through left and right animals' perceptual worlds. Brain Lang 2000; 73:189-219.
82. Deng C, Rogers LJ. Differential contributions of the two visual patways to functional lateralization in chicks. Behav Brain Res 1997; 87:173-182.
83. Deng C, Rogers LJ. Organisation of the tectorotundal and SP/IPS-rotundal projections in the chick. J Comp Neurol 1998; 394:171-185.
84. Deng C, Rogers LJ. Bilaterally projecting neurons in the two visual pathways of chicks. Brain Res 1998; 794:281-290.
85. Vallortigara G. Comparative neuropsychology of the dual brain: A stroll through left and right animals' perceptual worlds. Brain Lang 2000; 73:189-219.
86. Tommasi L, Vallortigara G. Encoding of geometric and landmark information in the left and right hemispheres of the avian brain. Behav Neurosci 2001; 115:602-613.
87. Deng C, Rogers LJ. Factors affecting the development of lateralization in chicks. In: Rogers LJ, Andrew RJ, eds. Comparative Vertebrate Lateralization. Cambridge: Cambridge University Press, 2002.
88. Güntürkün O. Avian visual lateralization: A review. Neuroreport 1997; 8:3-11.
89. Vallortigara G. Visual cognition and representation in birds and primates. In: Rogers LJ, Kaplan G, eds. Vertebrate Comparative Cognition: Are Primates Special? Kluwer Academic/Plenum Publishers, 2004.
90. Vallortigara G, Regolin L, Bortolomiol G et al. Lateral asymmetries due to preference in eye use during visual discrimination learning in chicks. Behav Brain Res 1996; 74:135-143.
91. Gross Y, Franko I, Lewin L. Effects of voluntary eye movements on hemispheric activity and choice of cognitive mode. Neuropsychologia 1978; 17:653-657.
92. Corballis PM, Fendrich R, Shapley RM et al. Illusory contour perception and amodal boundary completion: Evidence of a dissociation following callosotomy. J Cognitive Neurosci 1999; 11:459-466.
93. Ulrich C, Prior H, Duka T et al. Left hemispheric superiority for visuospatial orientation in homing pigeons. Behavioral Brain Res 1999; 104:169-178.
94. Vallortigara G, Pagni P, Sovrano VA. Separate geometric and nongeometric modules for spatial reorientation: Evidence from a lopsided animal brain. J Cogn Neurosci 2004; 16:390-400.
95. Vallortigara G, Regolin L. Facing an obstacle: Lateralization of object and spatial cognition. In: Andrew RJ, Rogers LJ, eds. Comparative Vertebrate Lateralization. Cambridge: Cambridge University Press, 2002:383-444.
96. Rogers LJ. The molecular neurobiology of early learning, development, and sensitive periods, with emphasis on the avian brain. Mol Neurobiol 1993; 7:161-187.
97. Rogers LJ. The development of brain and behavior in the chicken. Wallingford: CAB International, 1995.
98. Rogers LJ. Behavioral, structural and neurochemical asymmetries in the avian brain: A model system for studying visual development and processing. Neurosci Biobehav Rev 1996; 20:487-503.

CHAPTER 7

Lateralization of Spatial Orientation in Birds

Helmut Prior*

Abstract

Research on the specific role of the left and the right brain hemispheres during spatial orientation in birds is of great interest for several reasons. After it has become clear that lateralization is not restricted to humans, but is evolutionarily old and widespread among vertebrates, birds have established as one of the most important research models in this area. Furthermore, for a long time birds have been a major model in the study of spatial orientation in animals. In addition to behavior they share with vertebrates from other classes, they exhibit specific feats related to the fact that many species migrate and home over large distances. A number of recent studies show that, contrary to classical theory, spatial orientation in birds is not a mainly right-hemispheric task. It requires full involvement of either hemisphere, and several of the major orientation mechanisms are predominantly based in the left brain. During visual landmark orientation, both the left and right brain are essential, and they may contribute different aspects of spatial information. Mechanisms concerned with directional orientation strongly rely on the left brain hemisphere as olfactory orientation sun-compass based orientation and magnetic compass orientation. Comparative studies suggest a similar basic pattern of lateralization across species of different avian orders. However, along with different ecological niches this basic setup varies across species, leading to species-specific lateralization patterns for complex spatial behaviors.

Introduction

Research during the past two decades has shown that functional differences between the left and right brain hemisphere are widespread among vertebrates.[1-2] In birds, a large number of studies focused on visually guided behaviors and the underlying perceptive and cognitive abilities (reviews in refs. 1-3). In pigeons, chicks, and songbirds a left-hemispheric superiority for memorizing and discriminating object features has been found.[4-6] Regarding spatial information processing, early findings in chicks suggested a right hemispheric superiority.[2,7] Recent studies reveal, however, that the pattern of lateralization during spatial orientation in birds is more complex, and in the pigeon and in migrating songbirds left-hemispheric superiority or equal contribution of either hemisphere represents the typical finding. This has been demonstrated in the laboratory as well as in the field.

Research on lateralization of spatial orientation in pigeons began in 1997 with a field study that I carried out in cooperation with Onur Güntürkün, Hans-Peter Lipp, and the Swiss homing pigeon foundation.[8] There were three major aims. Firstly, until then there had been no studies at all on lateralization of spatial orientation during avian homing and migration. Secondly, all studies in birds had been carried out in rather small laboratory environments. Hence it was interesting to bridge the gap between the field and the laboratory. Thirdly, and most

*Helmut Prior—Allgemeine Psychologie I, Goethe-Universität Frankfurt am Main, Mertonstraße 17, D-60054 Frankfurt am Main, Germany. Email: Helmut.Prior@gmx.de

Behavioral and Morphological Asymmetries in Vertebrates, edited by Yegor B. Malashichev and A. Wallace Deckel. ©2006 Landes Bioscience.

importantly, there was a theoretical mismatch between hypotheses on the general function of the left and right brain hemisphere and hypotheses on domain-specific functions. Models on general processing principles in humans and other vertebrates suggested predominantly sequential information processing in the left brain and parallel processing in the right brain.[9] In the case of birds, this hypothesis is not consistent with the assumption that spatial orientation is generally a domain of the right hemisphere. For example, if birds are 'piloting' along landmarks or actively repeat sequences of snapshot-like scenes as revealed in chickens,[10] this clearly represents a case of sequential information processing. On the other hand, if a configuration of landmarks is used to determine a direction as it is suggested for homing pigeons at the release site (see ref. 11), this involves parallel processing of spatial information. Thus, either the theory on general processing principles or the hypothesis of a general right hemispheric specialization for spatial cognition in birds, or both, had to be revised.

Studies in Pigeons

In the first study we investigated lateralization in pigeons homing from familiar sites and along familiar routes.[8] As in most of our studies, we used the technique of monocular occlusion, and this investigation was the first animal study using this method in the field. In birds, fibres of the optic nerves cross over completely and a corpus callosum, the major connection between the left and the right forebrain of placental mammals, or other commissures of similar capacity are lacking. As a consequence, visual input to the right eye is mainly processed by the left brain hemisphere and vice versa. By temporarily covering one eye the performance of each brain hemisphere can be assessed separately.

In the training phase of the experiment no eyecaps were used. Thus, during becoming familiar with the three release sites (24-30 km from the loft) and routes the left and right hemisphere could compete freely for contribution to the learning process and to spatial memory storage. Subsequent monocular tests should reveal whether one of the brain hemispheres was more involved in the task than the other. Results showed considerably faster homing when the left hemisphere (right eye) was used. Figure 1 shows the homing speed for the last binocular release and the test release with the right or left eye covered. When birds used the right eye, homing times were similar to those in the binocular control condition. When birds used the left eye, homing was slower. A comparison of homing in sunshine and under overcast suggested that there were also hemispheric differences in directional orientation at the release site, but further testing was needed to confirm this. As an account for the observed functional asymmetry we suggested two alternative hypotheses. Firstly, it seemed possible that there is a left-hemispheric superiority in the visual memory for landscape features. Secondly, a left-hemispheric specialization for directional orientation was a likely candidate.

A follow-up study, carried out in cooperation with Rosie and Wolfgang Wiltschko, systematically assessed lateralized homing performance across five different release sites 40 to 50 km from the loft in Frankfurt am Main. In addition, the role of familiarity with the release site was addressed. Regarding homing times, findings confirmed those by Ulrich et al[8] though the effect was less marked. In addition, a profound lateralization of directional orientation at the release site was demonstrated. The classical measure of directional orientation is the vanishing bearing. When a pigeon is released, one or two observers follow the bird with glasses and note the compass direction where it could be seen last before it vanished from sight. In our study, performance with the left brain hemisphere was as good as in binocular controls (in spite of a systematic bias towards the side of the open eye), while it was poorer with the right hemisphere. A particularly interesting finding was that this lateralization of directional orientation emerged in the same way at familiar and at unfamiliar release sites. In order to test for the role of familiarity we used matched release sites, which had about the same distance from the loft, while the directions to the loft were opposite. Before the critical test releases, half of the birds were made familiar with one of the two release sites, the rest with the other release site. Then, birds were allocated to the test groups such that from each release site half of the birds knew the place and the other half of the birds were not familiar with it. As a measure of directional

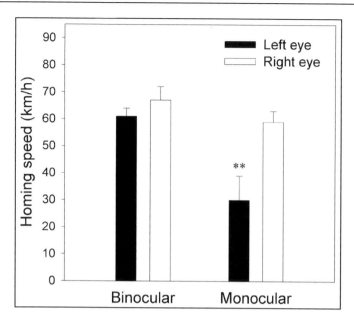

Figure 1. Homing speed in pigeons that had received binocular training flights and were then tested with the right or left eye occluded. When the left brain hemisphere (right eye) controlled behavior, performance was nearly as good as in the last binocular flight. With the right brain hemisphere (left eye), birds were considerably slower. **: p < 0.01, ANOVA. (Data from ref. 8).

orientation we calculated the angular difference, which is the absolute (regardless of the side) deviation from the control direction. An advantage of this measure is that it can be used in parametric statistical tests like an analysis of variance. Not surprisingly, birds that were familiar with the release sites showed, overall, smaller deviations. However, the pattern of lateralization was the same at familiar and unfamiliar sites. At familiar as well as at unfamiliar sites, the birds' orientation with the right eye only was as good as under binocular control conditions, but markedly poorer with the left eye (Fig. 2). Thus, the underlying neuronal processes were dependent on visual input, but did not require visual memory for landscape features.[5,12]

Further studies, which were conducted in cooperation with the group of Anna Gagliardo in Pisa in Tuscany, compared the performance from release sites within a short distance (10-15 km) under conditions that encouraged the use of visual cues (anosmic birds, familiar sites) or under conditions that prevented use of landmark cues at the release site (normal olfaction, unfamiliar sites). When homing was mainly guided by visual memory for landscape features, there was slightly faster homing with the left brain hemisphere,[13] but there was no difference in bearings at the release site. Without familiar landmarks, left-hemispheric superiority for orientation at the release site emerged.[14]

The findings from homing studies in pigeons with one eye occluded can be summarized as follows. There is a slight to moderate effect on homing times with superiority of the left-brain hemisphere. Lateralization of directional orientation depends on the environmental and experimental conditions. At visually familiar release sites, which are near-by (10 to 15 km), no lateralization occurs. A marked lateralization occurs when the release sites are farther away (40 to 50 km) and/or unknown. This lateralization does not depend on visual memory for landmarks.

Our findings with monocular occlusion are corroborated by results from lesion studies by other groups. There is left-hemispheric specialization for navigational map learning in pigeons with unilateral hippocampal lesions,[15] while both hemispheres contribute to familiar landmark navigation.[16]

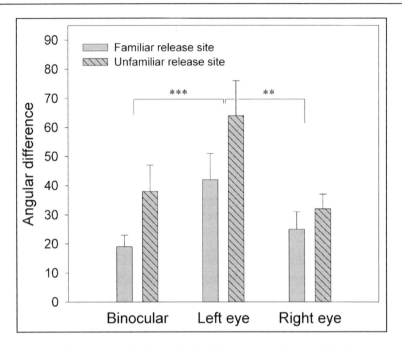

Figure 2. Directional orientation at familiar and unfamiliar release sites. Expectedly, birds were overall less directed at unfamiliar sites. In either condition, the same pattern of lateralization emerged: when the left brain hemisphere and right eye controlled the behavior (left eyecap), performance was equally good as in the binocular condition. When the right brain hemisphere controlled the behavior, orientation was poorer (data from ref. 5). ***: $p < 0.001$; **: $p < 0.01$. Experimental effects were analyzed by two-way ANOVA with the factors viewing condition and familiarity (both repeated measures). For details on the calculation of angular differences see text.

Furthermore, there is increasing evidence that the lateralization of performance in homing pigeons reflects a rather general left-hemispheric specialization in directional orientation. Two recent studies demonstrated a strong lateralization in favor of the left brain hemisphere for olfactory orientation[17] and for orientation based on the sun compass.[18]

In addition to homing studies in the field, we studied visuospatial orientation in semi-naturalistic laboratory settings. For working memory in a maze and navigation through a large indoor arena, similar patterns of lateralization emerged. In the arena study,[4] there were three sets of spatial cues: (1) global cues, in particular, the geometry of the room; (2) prominent landmarks within reach; and (3) small local cues that turned out to be irrelevant. As in the maze study, both hemispheres contributed equally to orientation by means of global visual cues, while there was left-hemispheric processing of object features and prominent landmarks within reach.[4,12] Figures 3 and 4 illustrate one of the critical tests from this arena study. Birds first had learned the path from a starting point (S) to a goal (G). They could have done so by relying on global room cues, by locating the goal with regard to the prominent landmarks, or by using both types of cues. In order to test, whether the right and left brain hemispheres differed in the use of cues, both types of cues were dissociated and the birds searched for the goal from two new starting points (A, B). The landmark array was translocated, keeping the relative positions of the landmarks to each other constant (Fig. 3). Now, there were two possible goals: One predicted by the global room cues (G), and the other (G'), predicted by the landmark array. In the beginning of the test session, the left and

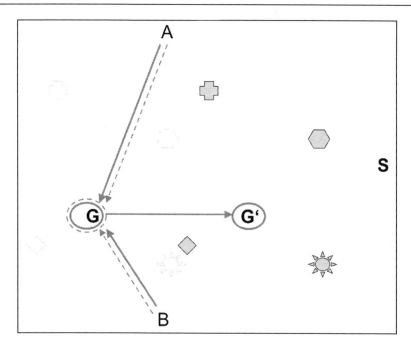

Figure 3. Behavior of pigeons with the right or left eye occluded after shift of the landmark array in a large laboratory arena. With the left brain hemisphere (right eye) as well as with the right brain hemisphere (left eye), subjects immediately went to the site predicted by global room cues, G, when searching from the new starting points A and B. With the left brain (solid arrows), they continued their search and went to the site predicted by the landmarks, G'. With the right brain (broken arrows), search was confined to the vicinity of G. Thus, the left brain hemisphere used both types of spatial information, while the right brain hemisphere only used the global cues. The symbols (cross, hexagon, diamond, sun) indicate the positions of the landmarks; on the left side (faded) during training, on the right side (bold) during the dissociation test.

the right brain hemispheres behaved similarly. In either condition, the pigeons directly approached the goal predicted by the global room cues within a few seconds (Fig. 3; Fig. 4, left), but then the behavior under the two conditions became strikingly different. With the left brain hemisphere (right eye) birds continued search and went to the position (G') predicted by the landmarks (Fig. 3). With the right brain hemisphere their search was confined to the vicinity of G (Fig. 3). Thus, for either brain hemisphere global room cues were most important. However, the left brain encoded and used both types of cues, while the right brain hemisphere only attended to the global room cues and completely ignored the landmarks. According with this difference in search behavior the search path length (p < 0.01, ANOVA, Fig. 4, right) as well as the search activity at G' (p < 0.01, Wilcoxon test) differed between the left and the right brain hemispheres.

Using a small laboratory arena and testing pigeons with left or right hippocampal lesions, Kahn and Bingman[19] found a higher reliance on global visual cues with the right brain intact than with the left brain intact. But also in this case, global visual cues were efficiently and predominantly used by either brain hemisphere.

Taken together, findings in pigeons suggest that both hemispheres contribute to a similar extent to orientation by means of global visuospatial cues. Orientation with single prominent landmarks is mainly performed by the left brain hemisphere. In addition, there are navigational mechanisms based in the left brain, which require further specification (see below).

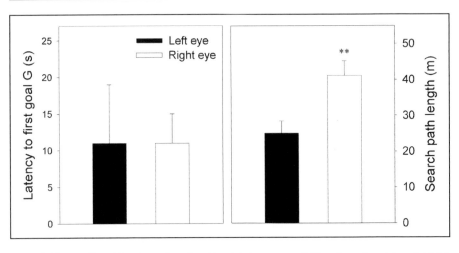

Figure 4. Time until reaching the goal predicted by global room cues (left) and search path length (right) in pigeons with the right or left eye occluded after shift of the landmark array in a large laboratory arena (see also Fig. 3). Under both conditions the goal predicted by the global room cues (G) was reached within the same time. But with the left brain hemisphere, birds continued to search for the goal predicted by the landmark array (G'). This is indicated by a longer search path. **: $p < 0.01$, ANOVA.

Studies in Migratory Birds

Parallel to our studies in pigeons we tested the hypothesis of a lateralized directional orientation in European robins (*Erithacus rubecula*), again in cooperation with Rosie and Wolfgang Wiltschko. In European robins, some of the birds are stationary, and others are migratory, in particular, those breeding in the north of the continent. We studied robins breeding in Scandinavia and spending the winter in the mild climate of western Germany. In this case, we were mainly interested in lateralization of orientation by means of magnetic information. Again, we used the technique of monocular occlusion. The effect of this might be twofold. In the first place, occluding one eye might simply determine whether the left or the right brain hemisphere takes control over behavior. There is, however, a more specific and intriguing possibility: several studies have suggested that the avian eye itself plays a role in magnetic compass orientation. At first glance, this appears quite surprising. But there is evidence from biophysics of photoreceptive biomolecules, that the magnetic field of the Earth can modulate photoreception in a way that depends on the angle between the magnetic field and the plane the photomolecules are arranged in. As a consequence, a specific modulation pattern is generated on the avian retina, with strong modulation of light perception at some and weaker modulation at other places.[20] Therefore, it is possible that occluding the right or left eye directly interferes with magnetoreception.

We caught the birds in mid winter at the Botanical Gardens of Frankfurt University. In late winter the birds were brought into migratory restlessness by shifting the daily light schedule to longer days. The birds' intended migratory direction was assessed in funnel cages with the magnetic field of the Earth as the only orientation cue. Funnel cages are equipped with a sensitive layer of paper, and every time a bird makes an attempt to fly it leaves a scratch mark on the paper. The distribution of scratch marks provides a reliable measure of the direction the bird wants to fly towards. The results from these experiments were clear-cut. Robins using the right eye and binocular controls were well-oriented into the direction of their Scandinavian breeding grounds (both: $p < 0.001$, Rayleigh tests). However, in robins using the left eye, orientation was at random.[21] This indicates a strong lateralization of magnetic compass orientation in favor of the left brain hemisphere (right eye). Replication of this finding in silvereyes

(*Zosterops lateralis*), a species of the southern hemisphere,[22] suggests that lateralization of magnetic compass orientation towards the right eye and left brain hemisphere is widespread among migratory passerine birds.

What is the proximate basis of this effect? As outlined above, the difference could be in the eyes as well as in central parts of the brain. If there is direct reception of the magnetic field with the eye,[20] photoreceptors in the avian retina should be crucial. So far, it is not known which photopigments are involved. As a likely candidate cryptochromes have been suggested because a specific retinal cell type, the displaced ganglion cell, is often associated with cryptochromes. Displaced ganglion cells have large receptive fields, making them suited to integrate visual input over larger areas of the retina. Therefore, the question arises whether there is an asymmetric distribution of cryptochromes in the retina of migratory birds. A recent study in garden warblers (*Sylvia borin*) did not find retinal asymmetries in the expression of one of the cryptochromes (CRY 1), although overall differences between garden warblers and nonmigratory controls were consistent with a role of CRY 1 in magnetoreception.[23] Cryptochromes have also been found in European robins.[24] In this species, for which contrary to garden warblers a behavioral lateralization of magnetic orientation is known, a possible retinal asymmetry of cryptochromes is currently investigated. So far, it is not clear whether between species differences are due to differences in retinal photoreceptors or due to differences in central processes. Further comparative studies are needed to clarify this.

Possible Lateralized Brain Mechanisms in Pigeons and Migratory Birds

For several reasons, lateralization of a central brain mechanism appears to be more likely than lateralization in peripheral sensory organs. Setting and representing a direction is a complex process. On theoretical grounds, optimizing the trade-off between efficient computing in specialized neural networks and efficient communication between networks involved with different aspects of complex tasks should lead to distributed networks and lateralization.[25-29] Although being of major importance for orientation, no study so far has specifically looked at brain mechanisms that could have the role of a direction setter. Aside from a mechanism that translates the input from the map, the compass and from motivational systems into the direction flown, there are several other systems that play a role in lateralization of directional orientation. A marked lateralization has been reported for olfactory orientation[17] and sun-compass based spatial learning.[18] For reasons of cognitive economy it seems likely that the directional information stemming from different sensory systems is processed by a single direction setter, which integrates multimodal input.

A Right-Hemispheric Specialization for Geometric Information?

Based on recent studies in chicks it has been suggested that geometric information may specifically be processed by the right brain hemisphere.[30] In studies on the use of geometric information, birds are tested in an arena that provides geometric information due to its shape (e.g., rectangular, triangular). As soon as the subjects have learned to orient within this arena, for example, during searching for food, the geometric information is varied (e.g., by changing the size) or put into conflict with other information (e.g., landmarks). Monocular testing can then reveal which role the geometric information plays for the left or right brain hemisphere. In the chick, there is clear evidence for hemispheric differences in the processing of geometric information. While either hemisphere makes use of this information, the left brain hemisphere is more attentive to the metric distance of walls and landmarks, while the right brain hemisphere is more concerned with the overall shape of the environment.[30] Regarding classical theory of avian spatial lateralization, it is worth noting that this finding does not support the hypothesis of right-hemispheric processing of geometric information. By contrast, it shows that both hemispheres are important for geometric information, and each is concerned with a specific aspect.

So far, studies in other avian species did not strictly focus on geometric information. For example, in their study with pigeons on navigation through a large indoor arena, which has been discussed above, Prior et al[4] separately assessed the role of prominent landmarks within reach and distant global cues. In this case, it is likely that geometric information played a major role, but a role of landmark-like global cues cannot be excluded. Kahn and Bingman[19] used a square arena so that geometric cues could not be used by the birds. Also, in the studies with food-storing and nonstoring passerine birds,[6,31-32] global cues comprised geometric and landmark-like aspects. In order to answer the question whether a similar hemispheric specialization for geometric information as in chicks is common in birds, studies focussing on strictly geometric information have to be carried out in the pigeon and in passerine birds.

Temporal Aspects

Findings in different paradigms of spatial representation and in different avian species in the majority of cases show essential involvement of either brain hemisphere or superiority of the left brain. When these findings are compared with those studies that support a predominant role of the right brain, it has to be considered that the studies differed in terms of the time window for spatial information processing. Experiments, which demonstrated most clearly left hemisphere superiority for remembering a location (feeding site, loft), concerned reference memory tasks. For example, in their studies with food-storing birds, Clayton and Krebs found a general left hemispheric superiority when they tested their birds one day after storing or later.[31-32] However, when the birds were tested after a retention interval of a few minutes,[6] the left hemisphere mainly focused on local object features while the right brain hemisphere relied on global room cues. In this study, birds could look for a food item, which was placed in one of several feeders distributed at different places in the experimental room. In addition to its specific position in the room each feeder had unique local features, e.g., a colour pattern. After finding the food item, birds then had to go to an adjacent room, and after a retention interval of a few minutes they were allowed to reenter the experimental room and search for the food. During the retention interval, the experimenter exchanged the feeder where the food had been with one of the other feeders. Thus, as in the arena experiment described above, there was a conflict between a site predicted by the global room cues and a site predicted by local cues. Thus, findings revealed a specific contribution of the right hemisphere after a short time interval, but not in the long-term. Together with other studies this suggests that the left avian brain hemisphere is the site where a stable long-term spatial reference is built up, while left and right brain hemispheres both play a role during initial temporary processing.

Regarding more general processing principles, it has been suggested that the right brain hemisphere is more sensitive to novelty and deals with a detailed physical representation, while the left brain hemisphere categorizes by means of stimulus features.[7] Thus, a possible scenario, which would resolve most of the discrepancies in the literature of the past two decades, is that during initial acquisition and real time processing of spatial information there is a complementary job sharing between the hemispheres, while an elaborate long-term representation is mainly stored in the left brain. Such a scenario is supported now by a large number of findings in several avian species. The model depicted in Figure 5 considers this aspect.

Comparative Lateralization of Avian Spatial Cognition

As shown above, global cues are the dominant cues for visuospatial orientation in pigeons, and they are efficiently processed by either brain hemisphere. Landmark information is processed by the left brain hemisphere in pigeons, and it is more or less neglected by the right brain.[4,19] In the chick, the situation is somewhat different in that processing of global visual cues is lateralized in favor of the right brain hemisphere. In both species and also in songbirds, strong lateralization of behavioral control is typically found if a conflict between different cues is created. This suggests that in either species both hemispheres are capable of processing spatial as well as object information. But in the pigeon, either hemisphere is highly competent in the orientation along global spatial cues, while information from prominent landmarks is mainly

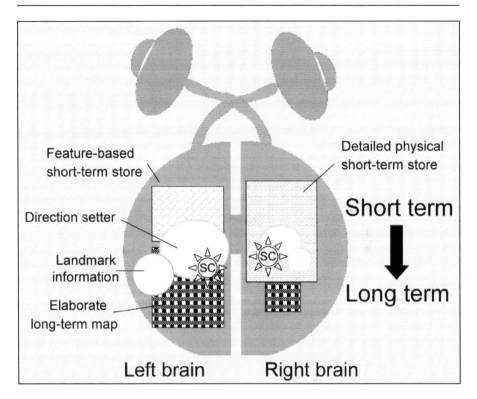

Figure 5. Schematic model integrating what currently is known about lateralization of spatial orientation in birds. During initial processing, the left hemisphere uses an elaborate, feature based encoding, while the right brain focuses on the physical aspects. An integrative long-term reference is predominantly built up by the left brain. A strong lateralization is found for directional orientation. A still tentative assumption of the model shown here is that the sensory systems (depicted here for the sun-compass) are bilateral and that the system setting the direction with input from different sensory systems is mainly based in the left brain. In addition, systems for landmark processing can be lateralized. The basic setup shown in the graph is modified depending on the species. For example, in the pigeon both hemispheres contribute to orientation along global visual cues, while the processing of prominent landmarks within reach is strongly lateralized. In the chick, an almost reverse pattern is found. SC: sun compass.

processed in the left brain.[4,19] In the chick, either hemisphere is good at object information but orientation along global spatial cues is mainly done with the right brain.[33]

When comparing species one should take care that the tests used and other conditions are as similar as possible. The interpretation of many findings in pigeons and passerine birds has been biased by generalizing from a few early findings in the chick, which has been proposed as the general avian model.[34] However, although being precocial, domestic chicks are still quite immature birds. And space use patterns differ greatly between adult chicken and adult birds of other species under study. It is unlikely that profound species differences as present between young chicks and adult pigeons should be paralleled by exactly the same neuronal organization and the same pattern of lateralization. Although adult chicken still can fly a little, they mainly move on the ground, while pigeons and migrating passerine birds regularly travel long distances high above the ground. Also, other aspects of spatial behavior and social behavior are strikingly different between pigeons and chicks. Although in groups of pigeons dominant and less dominant individuals occur, there is no strictly hierarchical organization as found in chickens.

Furthermore, until now there is no single study on lateralization of spatial orientation in adult chicken. In order to carry out valid species comparisons these data have to be collected. Therefore, a more definitive answer to the question which of the observed differences between the two best studied species are due to differences in testing age and which depend on true species differences, will only be possible when data on the lateralization in adult chicken have been obtained.

However, despite some inconsistencies in methods and data, a convergent comparative picture is emerging. Regarding the classic hypothesis of a general right hemispheric superiority,[35] which clearly is no more valid, it is of interest that recent experiments using a food-scanning task in a small laboratory environment demonstrated consistent evidence for more efficient scanning of the nearby area with the right brain in both, chicks and pigeons.[36] This might represent a rather specific phenomenon during search for food in the vicinity, but it also might reflect a more general superiority of the right brain for quick and efficient scanning of the surroundings. Such a hemispheric specialization for scanning could explain both a right hemispheric advantage for novelty detection and a right hemisphere advantage for global cue use in the short-term.

Summarizing the findings available and also considering the temporal aspects (see above), the lateralization model depicted in Figure 5 integrates what currently is known on the lateralization of spatial orientation in birds. This model can be summarized as follows. Both hemispheres are highly competent in the processing of global visuospatial cues, but they may focus on different aspects, particularly during initial processing. The initial percept of the left brain hemisphere is already elaborated while the percept of the right hemisphere is based on physical mapping. When recalling long-term spatial information, birds rely on an elaborate spatial map, which is mainly based in the left brain. In addition, there is strong lateralization of the left brain for directional orientation related to several sensory modalities.

Conclusion

Taken together, studies on lateralization of spatial orientation in birds show that this multi-component task is performed by both the left and right brain hemisphere. If one of the hemispheres plays a major role in avian spatial orientation, it is the left hemisphere in the majority of cases studied so far. For many aspects of spatial orientation, equal contribution of either hemisphere is essential for optimum performance, and the hemispheres carry out complementary subtasks. In cases where a stable, long-term reference is mainly built up in the left brain, the right hemisphere might nevertheless play an essential role as a temporary storage site of a detailed physical representation and as the site where quick and efficient scanning of the environment is carried out in order to detect novel and/or relevant stimuli.

Comparative approaches on spatial orientation and lateralization of spatial orientation in birds currently suffer from overgeneralization. As spatial behavior is rather diverse in avian species with different ecology, brain systems dealing with spatial orientation should show diversity that matches the behavioral diversity. It is reasonable to assume that specific spatial modules (e.g., compass systems) share a similar bauplan once they have evolved. The overall arrangement of brain systems dealing with spatial orientation, however, should show considerable variance across species. Rearrangement of lateralization patterns might play an important role in adjusting birds to different environments.

References

1. Vallortigara G, Rogers LJ, Bisazza A. Possible evolutionary origins of cognitive brain lateralization. Brain Res Rev 1999; 30:164-175.
2. Rogers LJ, Andrew RJ, eds. Comparative vertebrate lateralization. Cambridge: Cambridge University Press, 2002.
3. Güntürkün O. Avian visual lateralization: A review. Neuroreport 1997; 8:3-11.
4. Prior H, Lingenauber F, Nitschke J et al. Orientation and lateralized cue use in pigeons navigating a large indoor environment. J Exp Biol 2002; 205:1795-1805.
5. Prior H, Wiltschko R, Stapput K et al. Visual lateralization and homing in pigeons. Behav Brain Res 2004; 154:301-310.
6. Clayton NS, Krebs JR. Memory for spatial and object-specific cues in food storing and nonstoring birds. J Comp Physiol A 1994; 174:371-379.

7. Bradshaw JL, Rogers LJ. The evolution of lateral asymmetries, language, tool use, and intellect. San Diego: Academic Press, 1993.
8. Ulrich C, Prior H, Leshchinska I et al. Left-hemispheric superiority for visuospatial orientation in homing pigeons. Behav Brain Res 1999; 104:169-178.
9. Springer S, Deutsch G. Left brain, right brain. New York: Freeman, 1993.
10. Dawkins MS, Woodington A. Pattern recognition and active vision in chickens. Nature 2000; 403:652-655.
11. Bingman VP. Spatial representation and homing pigeon navigation. In: Healy SD, ed. Spatial Representation in Animals. Oxford: Oxford University Press, 1998:69-85.
12. Prior H, Güntürkün O. Parallel working memory for spatial location and food-related object-cues in foraging pigeons: Binocular and lateralized monocular performance. Learn Mem 2001; 8:44-51.
13. Diekamp B, Prior H, Ioaleé P et al. Effects of monocular viewing on orientation in an arena at the release site and homing performance in pigeons. Behav Brain Res 2002; 136:103-111.
14. Prior H, Diekamp B, Ioalè P et al. Lateralization of vanishing orientation in pigeons homing from unfamiliar release sites. Submitted.
15. Gagliardo A, Ioalè P, Odetti F et al. Hippocampus and homing in pigeons: Left and right hemispheric differences in navigational map learning. Eur J Neurosci 2001; 13:1617-1624.
16. Gagliardo A, Odetti F, Ioalè P et al. Bilateral participation of the hippocampus in familiar landmark navigation by homing pigeons. Behav Brain Res 2002; 136:201-209.
17. Gagliardo A, Odetti F, Ioalè P et al. Functional asymmetry of left and right avian piriform cortex in homing pigeon's navigation. Eur J Neurosci 2005; 22:189-194.
18. Gagliardo A, Vallortigara G, Nardi D et al. A lateralized avian hippocampus: Preferential role of the left hippocampal formation in homing pigeon sun compass-based spatial learning. Eur J Neurosci, (in press).
19. Kahn MC, Bingman VP. Lateralization of spatial learning in the avian hippocampal formation. Behav Neurosci 2004; 118:333-344.
20. Ritz T, Adem S, Schulten K. A model for vision-based magnetoreception in birds. Biophys J 2000; 78:707-718.
21. Wiltschko W, Traudt J, Güntürkün O et al. Lateralization of magnetic compass orientation in a migratory bird. Nature 2002; 419:467-470.
22. Wiltschko W, Munro U, Ford H et al. Lateralisation of magnetic compass orientation in Silvereyes, Zosterops lateralis. Australian Journal of Zoology 2003; 51:597-602.
23. Mouritsen H, Janssen-Bienhold U, Liedvogel M et al. Cryptochromes and neuronal-activity markers colocalize in the retina of migratory birds during magnetic orientation. Proc Natl Acad Sci USA 2004; 101:14294-14299.
24. Möller A, Sagasser S, Wiltschko W et al. Retinal cryptochrome in a migratory passerine bird: A possible transducer for the avian magnetic compass. Naturwissenschaften 2004; 91:585-588.
25. Kosslyn SM, Chabris CF, Marsolek CJ et al. Categorical versus coordinate spatial relations: Computational analyses and computer simulations. J Exp Psychol Hum Percept Perform 1992; 18:562-577.
26. Ringo JL, Doty RW, Demeter S et al. Time is of the essence: A conjecture that hemispheric specialization arises from interhemispheric conduction delay. Cereb Cortex 1994; 4:331-343.
27. Anderson B. Commentary. A proof of the need for the spatial clustering of interneuronal connections to enhance cortical computation. Cerebral Cortex 1999; 9:2-3, (Ringo, Doty, Demeter and Simard, Cerebral Cortex 1994; 4:331-343).
28. Levitan S, Reggia JA. A computational model of lateralization and asymmetries in cortical maps. Neur Comput 2000; 2:2037-2062.
29. Klyachko VA, Stevens CF. Connectiviy optimization and the positioning of cortical areas. Proc Natl Acad Sci USA 2003; 100:7937-7941.
30. Tommasi L, Vallortigara G. Encoding of geometric and landmark information in the left and right hemispheres of the avian brain. Behav Neurosci 2001; 115:602-613.
31. Clayton NS. Lateralization and unilateral transfer of spatial memory in marsh-tits. J Comp Physiol A 1993; 171:799-806.
32. Clayton NS, Krebs JR. Lateralization in paridae: Comparison of a storing and a nonstoring species on a one-trial associative memory task. J Comp Physiol A 1993; 171:807-815.
33. Regolin L, Garzotto B, Rugani R et al. Working memory in the chick: Parallel and lateralized mechnisms for encoding of object- and position-specific information. Behav Brain Res 2005; 157:1-9.
34. Rashid N, Andrew RJ. Right hemisphere advantages for topographical orientation in the domestic chick. Neuropsychologia 1989; 27:937-948.
35. Andrew RJ. The nature of behavioral lateralization. In: Andrew RJ, ed. Neural and Behavioral Plasticity: The Use of the Domestic Chick as a Model. Oxford: Oxford University Press, 1991:536-554.
36. Diekamp B, Regolin L, Güntürkün O et al. A left-sided visuospatial bias in birds. Curr Biol 2005; 15:R372-373.

CHAPTER 8

Lateralized Visual Processing in Anurans:
New Vistas through Ancient Eyes

Andrew Robins*

Abstract

The study of visual processing in anurans is of particular importance as the visual system of modern Amphibia is most similar to that possessed by the first tetrapods.[1] Anuran vision is one of the best studied sensory systems of all vertebrates, with both the hierarchical and integrative aspects of visual processing well described and extensively modeled. Nonetheless, current models of anuran visual processing fail to satisfactorily explain a range of behavioral functions.[2] This paper highlights key findings of lateralized visual processing in anurans to demonstrate the value of using such an approach to refine and expand on existing models. Social responses, predator-escape behaviors, predatory responses directed at familiar prey stimuli and the recognition of novel features of prey have been found to be under lateralized control by either the left or right side of the anuran brain. In each instance, the direction of visual lateralization mirrors that found in other vertebrate classes, particularly Aves and Mammalia. Such findings promote alternative avenues with which to further develop not only experiments in anuran visual processing, but also hypotheses regarding the conservation of behavioral functions specialized within the left or right sides of the vertebrate brain.

Introduction

'Lateralization' refers to the right and left sides or hemispheres of the brain being specialized to control different behavioral responses, or to process sensory information differently. Behavioral lateralization corresponds with neurological asymmetries found at both anatomical and neurophysiological levels.

A central hypothesis for the benefit of brain lateralization centres on the advantage provided by increases in the speed and efficiency of neural processes carried out by the respective hemispheres, resulting from a reduction in noise-resistance.[3-6] In other words, that the presence of a fully bilateral perceptual system could slow overall processing speed due to duplication and competition between the hemispheres. The duplication process may also be an expensive waste of metabolic resources. The ability to multi-task effectively and economically would be one major advantage of a lateralized brain: being able to guide feeding responses with one eye while monitoring the environment for a potential predator with the other eye is an often cited example from studies on the domestic chick.[6,7] Therefore, having a lateralized brain may give an advantage for species moving into and colonizing new environments. Such a challenge would have faced the first terrestrial tetrapods as they emerged from aquatic habitats: anurans provide excellent models with which to test this and other hypotheses as they are most closely representative of this early stage of evolution. Investigations of the presence, direction and strength of

*Andrew Robins—An independent researcher. Australia. Email: arobins@operamail.com

Behavioral and Morphological Asymmetries in Vertebrates, edited by Yegor B. Malashichev and A. Wallace Deckel. ©2006 Landes Bioscience.

visual lateralization in anurans provide an important comparative model for theories concerning the evolution of lateralized visual processing in the higher vertebrates.

This chapter will review studies of lateralized visual behavior in anuran amphibians. Speculations will then be made on candidate pathways for some lateralized visual functions, drawn from important findings across some 30 years of neuroethological research on the anuran visual system. I suggest that study of lateralized visual processing provides a new focus with which to consider issues of object-recognition and decision-making in these evolutionarily ancient animals.[2]

Comparative Visual Lateralization

In this review, 'lateralization' refers exclusively to behavior (measures of function), whereas 'asymmetry' relates mainly to neural morphology (measures of structure).[8] To avoid ambiguity, an individual possessing a behavioral bias for either side is said to have a 'preference' rather than 'individual lateralization'. Rather, lateralization refers exclusively to those behaviors where the majority of the group shares a consistent preference or 'sidedness' to perform a given response.[9]

Vertebrates posses a bilobar brain with each half controlling specialized functions (visual functions summarized in Fig. 1). The vertebrate brain is also under "crossed lateral control": sensory information from the left tends to be projected to the right side of the brain, which in turn controls the left side of the body, while sensory input and motor output on the right side tend to be integrated by the left side of the brain.[10] In vertebrates with optic nerves

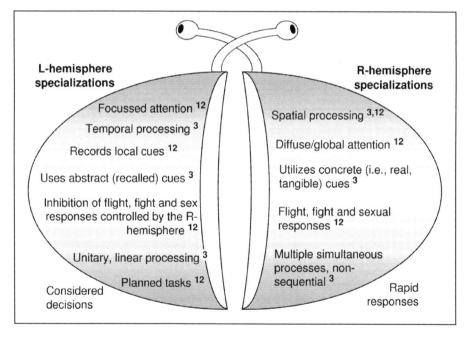

Figure 1. Comparative vertebrate lateralization. Specialized functions listed for either hemisphere have each been concluded from lateralized behavior observed in various species from at least two vertebrate classes (i.e., mammals, reptiles, birds, anuran amphibians and fish).[3,12] The lateralized functions are presented to highlight the complementary nature of processing in either brain hemisphere. The various functions are clearly associatively organized into those characterized by stored memories and deliberate, considered activities in the left hemisphere, and those functions in the right hemisphere requiring rapid responses to sudden or threatening changes in the environment. The pathways from the eyes are illustrated schematically to indicate the relative dominance of processing of visual information by either hemisphere (see text). (Data summarized from Bianki,[3] and Andrew and Rogers.[12])

projecting from either eye to both sides of the brain, dominance in processing and control of the behavioral response is directed by the hemisphere contralateral to the eye receiving the information.[5] Complete or near-complete decussation of the projections from laterally-positioned eyes to the contralateral sides of the brain occurs in bird, reptile, amphibian, and fish species.[11] Thus, the left and right eye systems (LES and RES, respectively) in the lower vertebrates can be considered as distinct left eye/right hemisphere and right eye/ left hemisphere pathways.

Vertebrate brain lateralization can be summarized at its most general form as follows. The right side of the brain is specialized to monitor the immediate surroundings for change.[12,13] Vigilance, the persistent attendance to both specific and general cues, is a related function of the right hemisphere.[13-15] Changes are detected and the appropriate behavioral response is physiologically matched by the right-hemisphere dominance in control of the fast-acting sympathetic nervous system.[16,17] (Although this relationship has only been examined in a few mammalian species: humans, dogs, cats, and rats, it is unlikely to have evolved de novo.) Thus, the right side of the brain is specialized for rapid responses to change or threat from the immediate surroundings, including social contexts, and can be considered chiefly associated with a state of high autonomic arousal.

By contrast to the functions of the right brain, the left side of the vertebrate brain is specialized for recording and recalling salient cues important for survival (see Fig. 1). Associated cues may be recalled and analysed in a step-wise, logical manner. The resultant mapping of previous experiences underpins the emergent functions of considered decision-making and manipulative motor acts (i.e., right-handedness for tool use) also associated with the left side of the brain.

The behavioral functions listed for vertebrates in Figure 1 suggest a general equivalence in perception and a corresponding level of cognitive or computational ability between the vertebrate classes. This is certainly not the case when making comparisons between the visual systems of amphibian and nonamphibian tetrapods, the former group characterized by considerable perceptual limitations, as the following sections show.

Ancestry of the Anuran Brain

The divergence of modern Amphibia from the lineage that eventually gave rise to mammals and the other tetrapods was estimated to have occurred around 300 mya (million years ago: Fig. 2).[18] Thereafter there were divergences of orders Apoda (or Caecilia: the burrowing, limbless and sightless amphibians) and Caudata (or Urodela: the Salamanders, characterized by secondary simplification of brain organization and of its genome to a paedomorphic state, where juvenile physical characteristics are retained in the mature stage of development).[19] Of the three orders, Anura remained, probably, closest to its Icthyostegan ancestry.

The amphibians are anamniotes, lacking the embryonic membranes (amnion and chorion), to protect the developing embryo from desiccation. Thus, the amphibians are developmentally more similar to the fish classes than to their amniotic tetrapod contemporaries. Nonetheless, while the anuran brain possesses the main brain divisions found in the higher vertebrates (i.e., olfactory bulbs, telencephalon, diencephalon, mesencephalon, cerebellum, and medulla oblongata: Fig. 3), their associated nuclei are less well defined than those found in advanced teleosts and sharks.[1,20]

The anuran brain has reduced metabolic requirements in comparison to its mammalian or avian counterpart, and while this factor reflects the wider physiological adaptations for survival in a wide range of environments that would kill other vertebrates, particularly mammals, it has been to the cost of the progressive evolution of the anuran brain.[1] The dependence on a 'cold-blooded' (poikilothermic) metabolism without the internal mechanisms regulating body temperature, has in part resulted in a high degree of phylogenetic conservation limiting both the size of the brain (relative to body size) and level of structural complexity.[1,20] Significantly, anurans lack the structural equivalent of the reptilian and avian cortex, or the mammalian neocortex,[21] and possess a correspondingly primitive visual system.

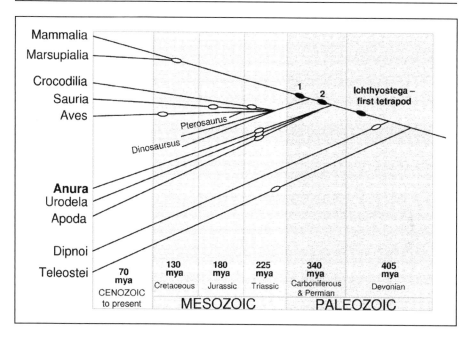

Figure 2. Vertebrate phylogeny. The relationship between vertebrates shown is based on fossil morphology and DNA data. Divergence times in mya (million years ago) are estimations only, with open pips showing earliest recognizable ancestors of the extant groups. Closed pips indicate further, comparatively radical divergences and evolutionary stages not otherwise shown in the mammalian line (1: Synapsidia, including the well-known *Dimetrodon*; 2: Cotylosauria, the first reptiles). Visual lateralization is yet to be investigated in urodelan and apodian orders of class Amphibia, and classes Crocodilia, Dipnoi, Chelonia, Ophidia and Coelacanthia (latter three classes omitted for clarity). Redrawn from Beçak and Kobashi. "Evolution by polyploidy and gene regulation in Anura." Genet Mol Res 2004; 3:195-212, by kind permission of Genetics and Molecular Research, http://www.funpecrp.com.br/gmr/index.htm, 2005.[18]

The anuran visual world is largely considered to be reduced to the detection of the edges and contours of objects in the surrounding terrain, and the movement of other stimuli as either potential 'predators' or 'prey'.[22-24] In a resting frog or toad, the retinal image of stationary objects is coarse and updated only when the animal moves its eye, head or body.[22,25] By contrast, any sudden changes or movements of discrete stimuli within the visual landscape of a motionless anuran are filtered and attended to selectively. The general properties of moving stimuli are analysed at multiple levels, classifying the stimuli before initiating the appropriate behavior (via a motor program generating system: MPS) as either 'escape predator' or 'approach prey'.[23,24]

Both the movement and the shape of the given stimulus is required to effect the appropriate response, the separate stimulus properties analysed together as a 'gestalt', or configuration.[23,26] To achieve such a level of perception, neural systems consisting of modifying and modulating feedback loops, and gating loops, are involved essentially simultaneously between different parts of the anuran visual system, as explained below.[24,27]

The Modular Visual System

The anuran retina lacks the equivalent of the avian or mammalian *fovea* or the reptilian visual streak for detailed visual analysis: most of the ganglion cell types are distributed evenly across the surface of the retina.[28] Nonetheless, the anuran retina is anatomically more complex than its mammalian counterpart and serves as the first stage of visual processing by operating as

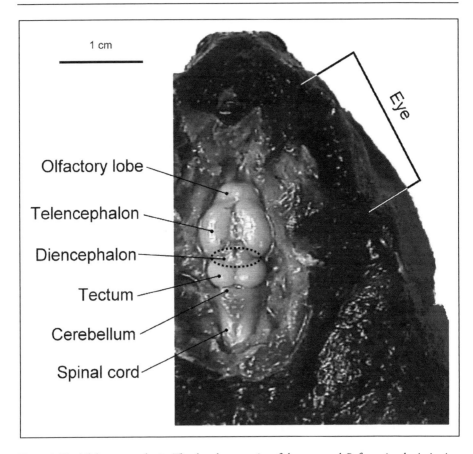

Figure 3. The bilobar anuran brain. The dorsal perspective of the cane toad *Bufo marinus* brain in situ. Structurally the brain is clearly divided although superficially symmetrical. The relative importance of the olfactory lobes, tectum, and eye is indicated by their size relative to the telencephalon, the structure in which comparatively higher processing is carried out. No cortical tissue is present in the amphibians, the cerebellum consists of a ridge of neural tissue immediately caudal to the tecta. The diencephalon is shown schematically as a deep area located beneath the telencephalon and tectum.

a dynamic filter.[26,29] In fact, the anuran ganglia carry out basic but selective filtering processes that, by contrast, are mainly carried out in the mammalian visual cortex.[26]

Essentially all retinal axons cross at the optic chiasma to project to the contralateral side of the brain.[28] Retinal projections to the anuran tectum are arranged topographically, and a columnar organization also exists which is comparable to that found in the mammalian visual cortex.[28] The tectum forms the dorsal roof of the mesencephalon, and is analogous to the mammalian superior colliculi in that it is principally involved with localization, assisting to orient the animal to specific stimuli.[23] The tectum continuously filters information from the retina and, without inhibitory inputs from higher centres, directs prey-catching behavior to all moving stimuli (see Fig. 4 also).[24,30]

The retinal projections to diencephalic (thalamic) centres are also organized topographically.[24] In the posterior thalamus, ganglion cell axons project to three regions collectively termed the praetectum (PT: Fig. 4). The main function of the PT is to "perceptually sharpen" specific features in the visual foreground and eliminating background cues.[2,27] The PT then specifies and controls the appropriate response to either 'predator' or 'prey' stimuli by appropriately

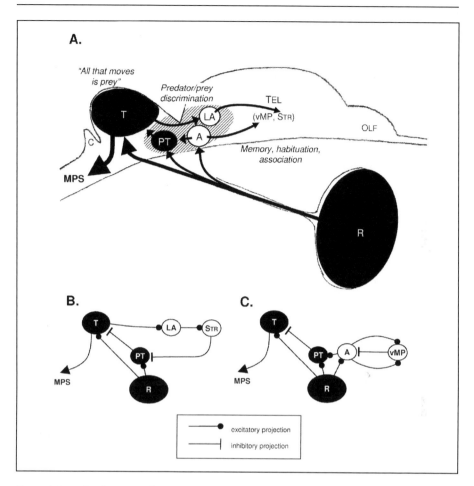

Figure 4. Visual information flow in the anuran brain. A) A lateral perspective of the main visual pathways in anurans, shown schematically for one side of the brain. Retinal ganglion cells project across the axial midline (not shown) to the tectum and diencephalon (shaded), but not to the telencephalon. The main functions served by the tectum, diencephalon and telencephalon are annotated in the figure. The tectum receives the majority of retinal projections and initiates visuomotor functions, particularly associated with orientation to prey, via motor programme generating systems. The tectum also projects back to higher centres in modulatory feedback loops shown in parts B and C. B) The stimulus-response gating loop model for mediating prey-catching responses in anurans.[27] The model shows the path by which the STR in the telencephalon is thought to inhibit the inhibitory responses of the PT on the tectum (T), thereby instigating prey-catching behavior (see text). C) The stimulus-response modifying loop for prey-catching responses.[27] The vMP in the telencephalon has been shown to play a central role in associative learning processes and habituation to familiar prey dummies (see text), and is proposed to modify the inhibitory activity of the PT via projections to A. Parts B and C modified from Ewert, Buxbaum-Conradi, Glagow et al. "Forebrain and midbrain structures involved in prey-catching behavior of toads: Stimulus-response mediating circuits and their modulating loops." Eur J Morphol 1999; 37:172-176, by kind permission of Taylor & Francis Ltd, http://www.tandf.co.uk/journals, 2005.[27] R: retina; Shaded region: diencephalon; PT: praetectal thalamic neurons (lateral posterodorsal thalamic nucleus, lateral posteroventral thalamic nucleus, and the posterior thalamic nucleus); A: anterior thalamic nucleus; LA: lateral anterior thalamic nucleus; TEL: telencephalon; STR: caudal striatum; vMP: ventral medial pallium; T: tectum; MPS: motor programme generating system; C: cerebellum; OLF: olfactory lobe.)

inhibiting or enhancing the prey-catching responses directed by the tectum.[22,24,31] Thus, objects with a rectangular shape moving continuously in the direction of the longitudinal axis are responded to as "worm" stimuli (potential prey) and elicits approach. However, the same shape moving perpendicular to the longitudinal axis in an "antiworm" configuration (i.e., moving upright, as if a snake in a strike position) tends to be responded to as a potential predator and elicits escape or defensive behaviors.[24,32] The activity of the anterior thalamus (AT: see Fig. 4) is mainly restricted to the frontal visual field, and inhibits prey-catching behaviors directed by neurons in the tectum.[24,28,33] The lateral anterior thalamic nucleus (LA) also receives ganglion afferents (not shown in Fig. 4), and is involved in higher-order functions in the telencephalon.[24,27]

Modulatory loops exist between the PT, AT, and LA regions of the diencephalon and the anuran telencephalon, a structure with which are attributed higher visual functions related to learning, memory and the recognition of novelty.[24] Two main areas associated with visual processing in the telencephalon are shown in Figure 4B,C: the caudal striatum (Str), and the ventral medial pallium (vMP). The Str plays a direct role in modulating the orienting responses of frogs and toads toward prey stimuli[34] and is thought to be involved with decision-making, controlling orienting responses mediated by the tectum in a process termed "striatal gating".[24,27,35] A direct ipsilateral striato-tectal descending projection is found in anurans and other amphibians but not in reptiles, birds and mammals (not shown in Fig. 4B).[34] These striato-tectal projections are thought to be inhibitory, whereas the indirect routes are thought to enhance prey-catching activity via the inhibition of otherwise inhibitory thalamo-tectal relays (see Fig. 4B).[24,27,34,36,37]

The vMP is involved in modulating the attentional state of the frog or toad.[34] Also known as the primordium hippocampi, the amphibian vMP is homologous to the mammalian hippocampus, and plays a major role in associative learning,[36,38,39] with particular regard to habituation to familiar prey and orienting responses towards novel prey stimuli (Fig. 4C).[27,40]

Experiments on Visual Lateralization in Anurans

Studies of prey-catching behavior in frogs and toads have provided neuroethology with a model system for visual pattern discrimination, as simple "sign" stimuli resembling potential predators or prey are used to generate specific and stereotypical behavioral responses.[22,26] The disassembling and reassembling of basic visual information is known for various retinal, thalamic and tectal neuron subtypes, however the manner in which such information is integrated in the face of complex visual stimuli is not well understood.[2] For example, predatory behavior in anurans (and other amphibians) is not as simple as earlier believed: the size and shape of the prey (i.e., worm versus insect) influences the decision to either bite or tongue-strike, and thus the predatory response is not confined to an inflexible chain of responses, or 'fixed action pattern' (FAP).[2] How are specific prey stimuli considered as their configuration is altered either by self-directed (i.e., movement patterns, defensive postures etc.) or predator-directed means (e.g., changing perspective on all or part of a "worm" stimulus)? What details of the prey are required to enable it to be recognized as noxious? The ways in which anurans discern and decide between available prey stimuli is unknown and although the factors of motivation and memory play an essential role, they are not easily incorporated into existing models of visual analysis.[2] The examples below show that the left and right sides of the anuran brain differ clearly in their perception of comparatively complex situations involving social cues, predator-detection, and prey stimuli. The following section then presents a series of hypotheses on how many lateralized responses may fit with established models of visual analysis in anurans.

Social Responses

Bisazza et al[41] made the first report of visual lateralization in larval anurans in social experiments performed on *Bufo bufo*, *B. viridis*, *Rana temporaria*, *R. esculenta* and *Bombina variegata* tadpoles. In an extensive battery of tests a left-eye/right hemisphere lateralization was concluded as individuals from each species oriented themselves with respect to mirrored surfaces.

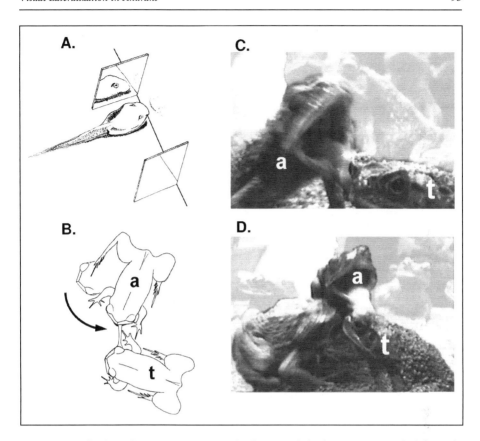

Figure 5. Lateralized social responses in anurans. A) When provided a choice to orient to the left or right side of a mirror, individual tadpoles of *Bufo bufo, B. viridis, Rana temporaria, R. esculenta,* and *Bombina variegata* inspect their mirrored image with left eye.[41] B) Toads (*B. bufo, B. marinus*) direct aggressive tongue-strikes preferentially at conspecifics located in their left visual hemifield. a: attacking toad; t: target toad. (Reprinted from Robins, Lippolis, Bisazza et al. "Lateralized agonistic responses and hindlimb use in toads." Anim Behav 1998; 56:875-881, ©1998 with permission from Elsevier.[43]) C) Example of an aggressive tongue-strike in *B. marinus* directed at the right eye of the target toad viewed in the attacker's right visual hemifield. D) Example of an aggressive tongue-strike in *B. marinus* directed at the right eye of the target toad viewed in the attacker's left visual hemifield. The same target toad is present in both Figures C and D, as are its attackers.

This lateralization is consistent with the right hemisphere specialization for social responses proposed for vertebrates in general.[12,41,42] Most testing involved individuals being placed in a modified aquarium with parallel and narrow mirrored surfaces positioned in a stall-like arrangement around the inside walls (see Fig. 5A). After 10 minutes, a significant side preference was noted for individual tadpoles to position themselves very close and comparatively closer to mirrors on their left.[41] No species differences were found in left side lateralization for either close or relative positioning to view the mirrors.

The first reports of lateralized social responses in adult anurans described aggressive tongue-striking behaviors in toads *Bufo marinus* and *B. bufo* (*B. viridis* were also tested but did not actively compete for the available prey and did not return sufficient scores).[43,44] Such strikes were directed at conspecifics located predominantly in the left visual hemifield of the attacking toad (see Fig. 5B-D).[43,44] The left-side lateralization for aggression towards conspecifics is not restricted to bufonid anurans as it has also been confirmed in sexually-immature tree frogs (*Litoria caerulea*: Robins and Rogers, paper in preparation). Left-eye lateralization for aggressive responses have been found in lizards (*Anolis carolinesis*,[45,46] *Urosaurus ornatus*,[47] and *Sceloporus virgatus*[48]), domestic chicks (*Gallus gallus domesticus*[49-51]) and gelada baboons (*Theropithecus gelada*[52]).

The main aspect of aggressive lateralization found in *B. marinus* was that significantly more strikes were directed at eyes to the attacker's left side: selective striking at the eyes of target toads occurred in both, but preferentially left, visual hemifields, suggesting that toads possess lateralized responses to specific visual stimuli as found in chicks. In 'social pecking tests', binocular chicks are more likely to peck at the heads (and not body or feet) of strangers in the left rather than the right lateral field of vision, whereas familiar companions are not pecked at preferentially in either the left or right lateral fields.[51]

Predator-Escape Responses

Lippolis et al[53] studied predator-escape responses in three bufonid species (*B. marinus*, *B. viridis*, and *B. bufo*) to a naturalistic predator model of a rubber snake head presented in a realistic two-stage movement pattern of slow approach followed by a rapid strike movement. No preference for jumping direction was found in the toad species tested when the snake model predator was introduced from the frontal binocular field.[53] However, for presentations of the snake model within the toads' lateral fields of vision, significantly higher levels of behavioral responsiveness were elicited when viewed with the toads' left eye than when viewed with the right eye (Fig. 6A-C). Lippolis et al[53] concluded that the visual lateralization found in the toads corresponded with the right hemisphere (left eye) specialization for spatial and affective behaviors found in other vertebrates.[12,54]

Rogers[55] studied individual *Litoria latopalmata* tadpoles and scored their turning preferences in two separate experiments. In the first experiment the turning preferences of individuals were tested in response to the sudden advance of a submerged vertical grid from directly in front of the visual midline. In tadpoles aged in Gosner[56] developmental stages 25-28 a significant although moderate lateralization to first turn right in 60-63% of cases was determined.[55] *L. latopalmata* tadpoles from stages 35-41 were reported as not responding to the advancing grid, but possessed right-turning lateralization in 61% of instances when escaping an advancing aerial predator model (Fig. 6D).[55] The rightwards lateralization may be attributable to motor biases alone, involving (yet unknown) asymmetries in the Mauthner cell complexes for fast or slow axial turning (found to be lateralized in various fish species[57]). However, it is most likely that this anti-predator response is the result of visual, and not motor lateralization: leftwards lateralization for turning rapidly after breathing air from the water's surface is found in larvae of many anuran species[58] including *L. latopalmata*,[55] a lateralization found in the absence of threatening visual stimuli. Thus, it appears likely that larval and adult anurans may share the left-eye mediated lateralization for predator-escape responses away to the right side.

Prey-Catching Responses

Standardized procedures for testing prey-catching responses in toads have been developed by Ewert and colleagues to reveal the specific functions carried out at different levels of the anuran visual system (including responses to "worm" and "antiworm" stimuli reviewed in a previous section).[23,32,59,60] Experimental paradigms based on the work of Ewert have recently been used to identify lateralized prey-catching responses in a range of toad species.[44,61] Essentially, individual anurans are contained within a transparent cylinder that is encircled continuously (clockwise or anticlockwise) by an automated 'dummy' prey item.[62-64] Individual members of the European toad species *B. bufo* and *B. viridis* were each tested once only with the dummy prey moving in either a clockwise or anticlockwise direction. The angular direction of tongue-strikes at the dummy prey (live, worm-like *Galleria mellonella* larvae tethered to cotton line) was scored from videotaped trials. The data from both bufonid species were analyzed together to reveal that significantly more strikes were directed at prey in the right hemifield than in the left hemifield.[44] Specifically, when the prey was moving clockwise, almost all strikes

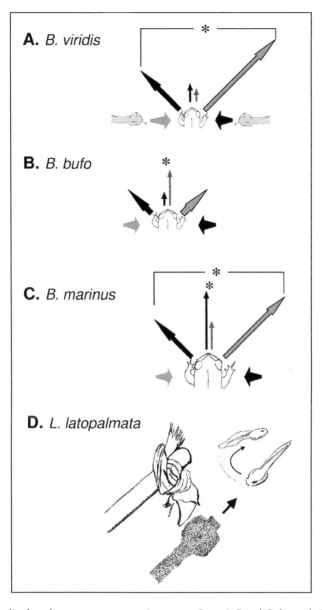

Figure 6. Lateralized predator-escape responses in anurans. Parts A, B and C show relative occurrence of frontal leaping (small arrows from toad's visual midline) and sideways leaping in response to a model predator introduced from the left or right lateral visual fields (a snake-head: shown in part A only). Across all three species tested Lippolis et al[53] concluded that the left eye was more reactive than the right eye to the predator stimulus. Note however that frontal leaping preferences in *B. marinus* (part C) suggests that the right and not left eye is dominant for mediating such behavior and indicates a species difference in reacting to the stimulus. Redrawn from Lippolis et al. "Lateralization of predator avoidance in three species of toads." Laterality 2002; 7:163-183, by kind permission of Psychology Press, www.psychologypress.co.uk/journals.asp, 2005.[53] D) *Litoria latopalmata* larvae are more likely to swim rightwards than leftwards during fast startle turning responses when threatened by an aerial predator model introduced from the frontal visual midline.[55]

occurred in the 'post-midline' area after the prey had crossed the midline from the left and moved into the right hemifield. This pattern of responses was not reversed for prey travelling in the reverse (anticlockwise) direction (see Fig. 7A,B).[44]

Previously published data of lateralized prey-catching in *B. marinus*[44] is expanded here to include the strike data from all toads tested (Fig. 7C), not only those toads observed making numerous aggressive responses in an earlier experiment.[43,44] *B. marinus* was found to direct significantly more strikes at dummy prey (tethered crickets) travelling clockwise and into the right hemifield, but no significant differences were found for any other comparisons.

A wider analysis of prey-catching responses in *B. marinus* was conducted in which both 'turns' toward, and 'strikes' directed at dummy prey were considered together as a single response.[61] The first experiment was the tethered cricket experiment of five minutes duration for which the 'strike' data was earlier presented (Fig. 8A, cf. Fig. 7C), here presented with the additional 'turn' data. The combined turn and strike data showed a lateralization consistent with the earlier assessment (cf. Fig. 7C), including significantly greater number of prey-catching responses in toads with prey moving clockwise than in anticlockwise directions (Fig. 8A).[61]

The second experiment examined prey-catching responses to a simple, wormlike horizontal strip (Fig. 8B) also conducted over five minutes of testing. No facilitation prey[32] were provided to the toads to promote predatory responses in the second experiment, in contrast to the first experiment in which a live cricket was placed in the cylinder with the toad to eat. The results provided a marked contrast to the data obtained from the previous experiment: the increase in the number of prey-catching responses elicited by the horizontal strip was generally an order of magnitude greater than those elicited earlier by the tethered crickets. Furthermore, no significant difference was found in the predatory responsiveness of toads presented with the horizontal strip moving clockwise or anticlockwise (Fig. 8B).[61] This result suggests that both eye systems respond with great rapidity to certain basic or 'key' aspects of prey stimuli. Such visual analyses and responses are mediated by the tecta,[62-64] concluded to be a nonlateralized centre of visual processing.[61]

No facilitation prey were provided to the toads in the first 5-minute period of the final experiment, for which the dummy prey was a black plastic insect (Fig. 8C). Similar to the experiment, in which a horizontal strip was used, in the first half of the trials no significant difference was found between the numbers of responses elicited by the model insect moving in either the clockwise or anticlockwise direction, in either left or right visual hemifield. Following the introduction of five mealworm larvae as facilitation prey at the start of the second five-minute period, lateralization was found for directing prey-catching responses at the dummy prey in the right visual hemifield, irrespective of its direction of movement (Fig. 8C).[61] The result was due to two effects: a marginal increase in receptivity to the model insect moving clockwise, and a marked suppression of receptivity to the same stimulus moving anticlockwise; the latter group tended to selectively ignore the dummy prey in favor of the mealworms.[61]

To summarize, prey stimuli with comparatively complex visual features (i.e., segmented body, appendages) were responded to comparatively slowly and revealed lateralized visual processing mediated by the right-eye (left side of the brain). This was shown most particularly as the toads were able to feed on facilitation prey, and indicated involvement of higher-level processing in the telencephalon due to associative visual, olfactory and gustatory cues.[24,36,38,39] The anuran telencephalon was therefore concluded to be lateralized.[61] Lateralized feeding responses preferentially controlled by the right eye (left hemisphere) have been found in chicks,[65-67] adult pigeons,[68,69] and zebra finches,[70] and reflect the specialization in the left hemisphere of the vertebrate brain for focused attention and considered responses (cf. Fig. 1).[12]

Responses to Novel Prey

To test the consistency of right-eye preferences in *B. marinus* for prey-catching, a choice-test was designed to test responses to stimuli viewed simultaneously, and exclusively, with either eye (Robins and Rogers, submitted). The test stimuli used had been viewed earlier by the toads; they were black insect models identical to that used as dummy prey in the previous experiment. Two

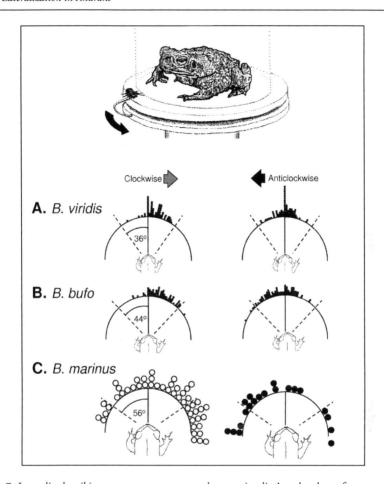

Figure 7. Lateralized striking responses at automated prey stimuli. Angular data of tongue-striking frequency and location in response to a single prey dummy travelling either clockwise or anticlockwise about an individual toad contained in a transparent cylinder. (Figure inset, reprinted from Robins and Rogers. "Lateralized prey catching responses in the toad (*Bufo marinus*): analysis of complex visual stimuli." Anim Behav 2004; 68:767-775, ©2004, with permission from Elsevier.[61]) For A, *B. viridis* and B, *B. bufo*, the dummy prey used were tethered live *Galleria mellonella* larvae moving at one revolution per minute (rpm) for two minutes only. A general lateralization to strike at the prey in the right visual hemifield was found for *B. viridis* and *B. bufo*, an effect strongest with the dummy prey moving in a clockwise direction.[44] C) *B. marinus* toads were tested using tethered live adult crickets (*Acheta sp.*) moving at 1.7 (rpm) during five-minute trials, and are presented differently to illustrate these procedural differences. Despite the testing differences, *B. marinus* was found to direct significantly more strikes at dummy prey travelling clockwise and into the right hemifield (13 toads each were tested with the dummy prey moving either clockwise or anticlockwise. Wilcoxon Mann-Whitney U-test, two-tailed: $W = 124.00$, $Z = -2.26$, $p = 0.024$). The strike location data for the toad species in Parts A and B have been presented in Vallortigara et al. "Complementary right and left hemifield use for predatory and agonistic behavior in toads." Neuroreport 1998; 9(14):3341-3344, with kind permission from publishers Lippincott Williams and Wilkins.[44] The figures are here modified to include the area of binocular overlap for the respective species.[53] Note the tendency for a decrease in strikes along or close to the visual midline, with a comparative increase in strikes in the lateral monocular fields, commensurate with the frontal positioning of the eyes in each of the anuran species. Further experiments are necessary to confirm this pattern as an effect of prey-catching specializations corresponding with species differences in the binocular field overlap and the positioning of the eyes.

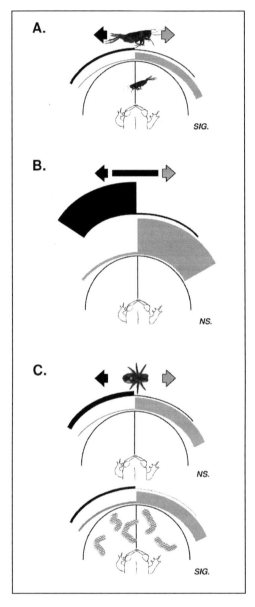

Figure 8. Lateralized prey-catching responses (turns toward and strikes) in *B. marinus*. The results from a series of trials using toads tested in five-minute trials once only with the dummy prey moving at 1.7 rpm anticlockwise (black arrow) or clockwise (grey arrow). The relative frequency of responses are shown as widths of distribution bars (anticlockwise: black; clockwise: grey bars) to dummy prey viewed in either left or right visual hemifield. A) Live tethered cricket as dummy prey with an additional cricket provided to facilitate responsiveness (see text) reveals a lateralized preference for dummy prey moving clockwise into right visual hemifield. B) Horizontal black strip (no facilitation prey) elicits high numbers of predatory responses responsiveness, no lateralization revealed. C) Plastic black insect used as dummy prey in two five-minute blocks. No facilitation prey provided in first block, no lateralization revealed. Five mealworm larvae provided to facilitate predatory responsiveness of toads in second block, lateralization for dummy prey in right visual hemifield revealed. Summary figures constructed from data provided in reference 61.

test stimuli were attached to the end of a wire 'Y'-fork and introduced from behind the toad as it fixated on a computerized 'zeroing' stimulus moving on a monitor screen. Although the shape of the test stimuli were known to the toads, their configurational properties were novel, resembling hovering insects at eye-height as they were manually vibrated in repeated presentations.

Somewhat unexpectedly, the toads first turned to inspect the test stimulus in the left, and not right, lateral visual field in the significant majority of cases. This lateralization was lost for those individuals turning to inspect the unpalatable test stimuli more than once, and habituation to successive presentations of the test stimuli was rapid. No lateralization was found when the toads were retested on subsequent occasions, until the toads were retested with a novel white stripe painted on the dorsal thorax and abdomen of the black insect models. Only then was the lateralization for making the first turn to the test stimulus in the left lateral field was reestablished. Thus, responses to the black insect stimuli seen hovering for the first time, and to the hovering insect painted with a white stripe were the only instances in which any lateralization (both leftwards) was observed.

The left side bias (right hemisphere specialization) in toads for responding to novel stimuli is consistent with the left eye (and right hemisphere) specialization (LES) for detecting novel stimuli in the chick.[65,66,71,72] More specifically, the LES of the chick is concerned with the analysis of all properties of an object or experience, and is specialized to detect or attend to unpredictable and novel changes to a given object or environment.[71,73,74]

New Models from Old Modules

Figure 9 presents a series of previously-published models regarding visual analysis in the anuran brain, drawn here illustrating candidate telencephalic centres likely to be involved with lateralized responses to predators, familiar and novel prey stimuli. For brevity, the pathways are presented for the LES, acknowledging also that pathways in the RES are mirror-imaged in their general organization. Putative asymmetries in the morphology of the visual centres, projecting axon diameter and number of fibres, etc. are shown highlighted, reflecting the respective forms of visual lateralization.

Lateralized Predator-Escape Responses

Asymmetries associated with the Str in the right telencephalon (LES) may be correlated with lateralization of predator escape responses in anuran species (Fig. 9A). Striatal gating for predator stimuli is suggested from the inhibitory striato-tectal projection,[34] in association with retinal projections to the LA.[24] These processes may be modulated by a feedback loop via the PT and tectum, a somewhat longer route postulated for gating appropriate responses to prey stimuli (and also predator stimuli, by elimination: see Fig. 9A).[24,26,27,35,40] Noteworthy also is the central role of the Str in recalling the spatial location of barriers and escape routes when evading threatening stimuli (*Rana pipiens*),[22] and for localizing the source of water-waves indicating the presence of invading conspecifics to aquatic territories (*Xenopus laevis*).[75] Thus, the anuran Str appears to integrate spatial information (from different modalities) in a manner that suggests a role for monitoring the visual background for sources of potential threat—functions which are lateralized within the right hemisphere of higher vertebrates (cf. Fig. 1).[12,13]

Lateralized Prey-Catching Responses to Familiar, Complex Stimuli

Increases in glucose utilization by the vMP occurs following habituation to a familiar prey dummy, a process thought to correspond with the storage of associated cues within the vMP to form a spatiotemporal 'model' of the familiar prey.[27,40,76] The formation of such a model is further thought to involve the ipsilateral Str, in addition to cross-modal inputs, in a manner analogous to associative memory formation in the mammalian hippocampus.[40,76] Thus, the vMP is hypothetically involved with 'match' and 'mismatch' analyses comparing the currently viewed prey stimulus with previous experience. 'Matching' to familiar prey promotes habituation and suppresses approach to the prey, while 'mismatching' to novel prey reliably elicits approach.[40]

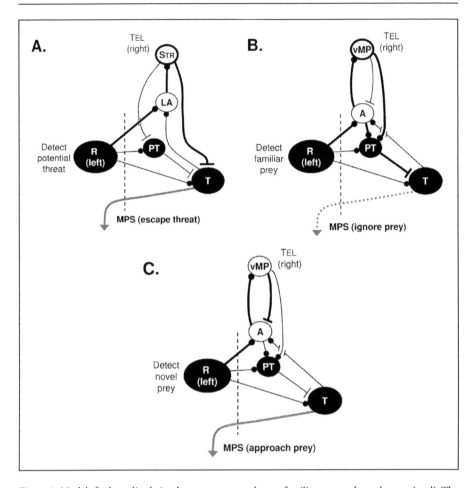

Figure 9. Models for lateralized visual responses to predators, familiar prey and novel prey stimuli. The three models are extensions from those postulated by Ewert and coworkers for prey-catching responses (e.g., refs. 24,26,27,40,76). A) Mechanism for lateralized responses to predator stimuli, utilizing modifications of the "striatal gating loop" (see Fig. 4B). Postulated asymmetries correlating with the dominance in detecting and/or responding to threatening stimuli viewed by the left eye of anurans are shown highlighted. A major component of the lateralization is suggested for the right striato-tectal route and its inhibitory role in suppressing orientation towards the novel stimulus.[34,37] B) Mechanism for lateralized habituation to familiar prey by the right telencephalon. The modulating loop between the right vMP and the right PT enhances the inhibitory role on the prey-catching responses of the tectum (T), whereby familiar prey is "matched" to a stored representation (see text). The process of habituation is driven by the right vMP in conjunction with associative learning processes via the ipsilateral T and A (e.g., refs. 24,27,40), in addition to spatiotemporal input from the ipsilateral STR (not shown: see ref. 27). C) Mechanism for lateralized responses to novel prey stimuli. The consequence of a "mismatch" between past and present prey items generated by the right vMP is shown highlighted, inhibiting the ipsilateral A and modulating the inhibitory role of the PT. Novel prey items viewed with the left and not right eye are preferentially investigated. Figures modified from Ewert, Buxbaum-Conradi, Glagow, et al. "Forebrain and midbrain structures involved in prey-catching behavior of toads: Stimulus-response mediating circuits and their modulating loops." Eur J Morphol 1999; 37:172-176, by kind permission of Taylor & Francis Ltd, http://www.tandf.co.uk/journals, 2005.[27] R: retina; PT: praetectal thalamic neurons; A: anterior thalamic nucleus; LA: lateral anterior thalamic nucleus; TEL: telencephalon; STR: caudal striatum; vMP: ventral medial pallium; T: tectum; MPS: motor program generating system.

The basis for the dominance of the RES for prey-catching responses at familiar prey stimuli (cf. Figs. 7,8) is indirect: it is assumed here that the RES habituates slower than the LES to the same stimuli, the latter involving the central role of the right vMP (the habituation pathway modelled in Fig. 9B). Presumably the rapid categorization of the prey stimuli as 'familiar' and nonthreatening by the right vMP then frees valuable resources in the right telencephalon (i.e., right Str) to monitor and attend other stimuli. The superiority of the right vMP for memory of, and habituation to, familiar prey, is the most parsimonious explanation to the prey-catching data where the RES (left vMP) shows demonstrable persistence and not habituation to the familiar dummy prey. The apparent division of function between the left and right vMP is particularly evident when facilitation prey provides associated olfactory, visual and gustatory inputs (cf. Fig. 8C).

Lateralized Responses to Novel Prey Stimuli

Figure 9C shows the pathways activated when a novel prey stimulus is 'mismatched' with memories of familiar prey stored in the right vMP, in contrast to the model described above for right vMP 'matching' to familiar prey (see Fig. 9B). Lateralized responses to novel prey mediated by the LES ensue. By corollary, it is thought that the right hemisphere of the chick also holds detailed records of complex stimuli, and uses these records to detect 'novelty'.[71]

Discussion

The left and right sides of the anuran brain clearly posses contrasting motivational agendas: they respond differently to the same visual cues relating to conspecifics, predators and prey. The anuran brain hemispheres are not homogenous, as historically assumed in neuroethological research. More specifically, the anuran telencephalon possesses lateralized processing for visual information, there is no evidence yet indicating the tectum to be similarly lateralized. Future unilateral antidromic-recording and ablation studies of the anuran visual system should report the side of the brain under examination. This small but significant step would assist substantially in replication studies in anuran neuroethology, and aid in the interpretation of responses to specific visual stimuli of greater complexity than simple sign stimuli.

Whether or not visual lateralizations correlate with asymmetries in the telencephalic structures, as suggested in models presented in the previous section, remains to be tested. Asymmetries in the visual system may be present not only in higher telencephalic centres but also in the diencephalon (i.e., lateral anterior thalamic nucleus, anterior thalamic nucleus, and praetectum) and mesencephalon (i.e., tectum, nucleus isthmi, and antereoventral tegmenti). Similar asymmetries are found in multiple levels of the avian visual system.[77-79] Irrespective of the locations of structural asymmetry, the behavioral laterality identified in anurans corresponds directly with those, where known, found in higher vertebrates and fish. It is considered unlikely that all four lateralized visual processes reviewed in this chapter (see Fig. 10) result from convergent evolution in the mammalian, reptilian, avian and anuran lines. The homology for social responses, rapid prey-escape responses, considered prey-catching responses, and responses to novel prey is all the more remarkable given considerable differences in the organization of the primary visual systems of anurans and the rest of the vertebrate subphyla.

One important aspect from studies of laterality of visual processing in anurans is that the degree of lateralization is modifiable with experience. Examples include the lateralization of turning responses in subsequent presentations of threatening stimuli in larval anurans and novel prey stimuli in adult anurans. The loss of lateralization infers interhemispheric transfer and memory of particular aspects of the visual stimulus. That is, the process of interhemispheric transfer abrogates to some degree the apparent flaws in possessing a lateralized brain: that a brain hemisphere and eye-system specialized for some functions is deficient for others. Thus, the value that a lateralized brain has for driving behavioral adaptation and colonizing novel environments more than offsets the laterality for detecting and avoiding predators, a deficiency that could be exploited by predators themselves.

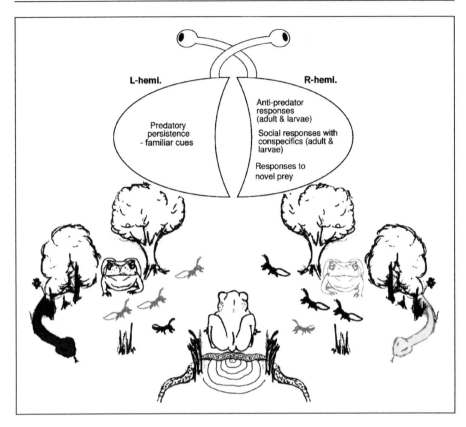

Figure 10. What the toad's lateralized brain sees from the toads eyes. Four main types of lateralization have been identified in anurans, summarized in this figure to illustrate that anurans do not possess perceptual systems mirror-imaged in both brain hemispheres. The left eye/right hemisphere appears to have greater sensitivity to threatening stimuli than its opposite counterpart, and this is true also of social stimuli for either aggressive or grouping behavior. Novel prey stimuli are also investigated preferentially by the left eye/right brain hemisphere. By contrast, familiar prey, or stimuli presented within familiar or nonthreatening environs, are attended to with greater persistence with the right eye/left brain hemisphere. This complementary specialization may be related to the important roles of the left eye/right brain hemisphere for monitoring the surrounding environment for potentially threatening changes.

It is difficult to envisage more important and mutually exclusive functions than 'avoid predator' and 'detect prey' that would be advantaged by complementary compartmentalization into right and left brain hemispheres. The lateralization of these core functions may well have formed the foundation for innovative and emergent features (e.g., lateralization of spatial processing, social behaviors including communication and language). The level of sophistication attained in lateralized behaviors then reflect the manner in which the vertebrate brain has been elaborated upon in successive stages with the formation and lamination (often independently evolved) of cortical tissue, and the lamination of sub-cortical tissues.[21] It is arguable that a perhaps conclusive test of the importance of conserved forms of visual lateralization to prey and predators is to examine urodelan amphibians (i.e., salamanders and newts) for lateralized responses to such stimuli. As the brains of members of the Urodela have undergone secondary simplification in the process of divergence from the anuran line,[19] the retention or absence of visual lateralization in urodelans would have great importance for hypotheses on the conservation of vital behavioral responses.

The contrasting behavioral responses directed by the left and right eye systems reviewed in this chapter provide a fresh perspective with which to address the important issues of visual discrimination, learning and memory in anurans and other lower vertebrates. In the brief range of experiments of lateralized behavior conducted so far, insights have been made regarding detailed object recognition and complex decision-making. This is particularly significant as these cognitive behaviors have been concluded from vertebrates lacking the neural complexity otherwise normally associated with such levels of cognition. While the assumption of an essentially homologous evolution of brain lateralization in vertebrates is strengthened by the data from anuran behavior, there are still crucial confirmatory experiments to be conducted. One such area currently overlooked is whether anurans possess right-hemisphere lateralization for spatial processing: an attribute important in territorial species, and subject to seasonal and hormonal fluctuations. In a related function, it is yet to be confirmed that anurans possess left-hemisphere dominance in vocalization although an early report suggests this to be the case, as found in higher vertebrates.[79] In a similar vein, it is essential to confirm whether the anuran (and amphibian) autonomic nervous system is under lateralized control. Of specific interest is whether the direction of lateralization matches that found in higher vertebrates, with the right brain predominately controlling the sympathetic nervous system and the left brain predominately controlling the parasympathetic nervous system. If so, the laterality of these ancient regulatory systems will correspond well with the cognitive roles of the respective brain hemispheres: the sympathetic nervous activity matched to vigilance functions mediated by the right hemisphere and parasympathetic nervous activity conducive to the consolidation of long-term memories and learned responses mediated by the left hemisphere. Is this relationship merely correlated or does laterality of the autonomic nervous system form a compelling basis for the evolution of lateralized cognitive processes? In the light of these and similar future investigations, it is not implausible to suggest that anuran laterality provides an important avenue with which to model the effects of environmental constraints on the evolution of the vertebrate brain, and perhaps also the roots of cognition themselves.

Acknowledgements

A. Robins wishes to thank Giuseppe Lippolis for his preparation of tissue used in Figure 3, and also for his concept for the summary diagram of Figure 10. The help of Lesley J. Rogers is also acknowledged and warmly appreciated for her guidance regarding A. Robins' doctoral research project conducted at the University of New England, Australia: many of the publications of lateralization in anurans discussed in this review stem from that research. Thanks also for the marvellous organization and commitment of Yegor Malashichev, St. Petersburg State University, for getting 2-ISBMA up and running and bringing this compilation to fruition. Love and thanks to my wife, Megan, for supporting my journey to St. Petersburg.

References

1. Wilczynski W. The nervous system. In: Feder ME, Burggren WW, eds. Environmental Physiology of the Amphibians. Chicago: University of Chicago Press, 1992:9-39.
2. Roth G, Dicke U, Wiggers W. Vision. In: Heatwole H, Dawley EM, eds. Amphibian Biology. Chipping-Norton: Surrey Beatty & Sons, 1998:783-877.
3. Bianki VL. The Right and Left Hemispheres of the Animal Brain: Cerebral Lateralization of Cerebral Function. Monographs in Neuroscience. New York: Gordon and Breach, 1988.
4. Andrew RJ. The nature of behavioral lateralization in the chick. In: Andrew RJ, ed. Neural and Behavioral Plasticity: The Use of the Chick as a Model. Oxford: Oxford University Press, 1991:536-554.
5. Bradshaw JL, Rogers LJ. The Evolution of Lateral Asymmetries, Language, Tool Use, and Intellect. New York: Academic Press, 1993.
6. Rogers LJ. Evolution of hemispheric specialization: Advantages and disadvantages. Brain Lang 2000; 73:236-253.
7. Rogers LJ. Advantages and disadvantages of lateralisation. In: Rogers LJ, Andrew RJ, eds. Comparative Vertebrate Lateralization. Cambridge: Cambridge University Press, 2002.
8. Rogers LJ. Laterality in animals. Int J Comp Psychol 1989; 3:5-25.

9. Rogers LJ. Evolution of side biases: Motor versus sensory lateralization. In: Mandal MK, Bulman-Fleming MB, Tiwari G, eds. Side Bias: A Neurophysiological Perspective. Amsterdam: Kluver Academic Publishers, 2000:3-40.

10. Walker SF. Lateralization of functions in the vertebrate brain: A review. Br J Psychol 1980; 71:329-367.

11. Bisazza A, Rogers LJ, Vallortigara G. The origins of cerebral asymmetry: A review of evidence of behavioral and brain lateralization in fishes, reptiles and amphibians. Neurosci Biobehav Rev 1998; 22:411-426.

12. Andrew RJ, Rogers LJ. The nature of lateralization in tetrapods. In: Rogers LJ, Andrew RJ, eds. Comparative Vertebrate Lateralization. Cambridge: Cambridge University Press, 2002:94-125.

13. Rogers LJ. Lateralization in vertebrates: Its early evolution, general patterns and development. In: Slater PJB, Rosenblatt J, Snowdon C, Roper T, eds. Advances in the Study of Behavior. San Diego: Academic Press, 2002:107-162.

14. Heilman KM. Attentional Asymmetries. In: Davidson RJ, Hugdahl K, eds. Brain Asymmetry. Cambridge, Massachusetts: MIT Press, 1995:217-233.

15. Cherry BJ, Hellige JB. Hemispheric asymmetries in vigilance and cerebral arousal mechanisms in younger and older adults. Neuropsychology 1999; 13:111-120.

16. Lane RD, Jennings JR. Hemispheric asymmetry, autonomic asymmetry, and the problem of sudden cardiac death. In: Davidson RJ, Hugdahl. K, eds. Brain Asymmetry. Cambridge, MA: MIT Press, 1995:271-304.

17. Wittling W. Brain asymmetry in the control of autonomic-physiologic activity. In: Davidson RJ, Hugdahl K, eds. Brain Asymmetry. Cambridge, Massachusetts: MIT Press, 1995:305-357.

18. Beçak MA, Kobashi LS. Evolution by polyploidy and gene regulation in Anura. Genet Mol Res 2004; 3:195-212.

19. Roth G, Nishikawa KC, Wake DB. Genome size, secondary simplification, and the evolution of the brain in Salamanders. Brain Behav Evol 1997; 50:50-59.

20. Taylor GM, Nol E, Boire D. Brain regions and encephalization in anurans: Adaptation or stability? Brain Behav Evol 1995; 45:96-109.

21. Striedter GF. The telencephalon of tetrapods in evolution. Brain Behav Evol 1997; 49:179-213.

22. Ingle DJ, Hoff KVS. Visually elicited evasive behavior in frogs. Bioscience 1990; 40:284-291.

23. Ewert JP. Neural mechanisms of prey-catching and avoidance behavior in the toad (Bufo bufo L.). Brain Behav Evol 1970; 3:36-56.

24. Ewert JP. Neural correlates of key stimulus and releasing mechanism: A case study and two concepts. Trends Neurosci 1997; 20:332-339.

25. Grüsser OJ, Grüsser-Cornehls U. Neurophysiology of the anuran visual system. In: Llinás R, Precht W, eds. Frog Neurobiology. Berlin: Springer-Verlag, 1976:297-385.

26. Ewert JP. Neuroethology of releasing mechanisms: Prey-catching in toads. Behav Brain Sci 1987; 10:337-405.

27. Ewert JP, Buxbaum-Conradi H, Glagow M et al. Forebrain and midbrain structures involved in prey-catching behavior of toads: Stimulus-response mediating circuits and their modulating loops. Eur J Morphol 1999; 37:172-176.

28. Gaillard F. Visual units in the central nervous system of the frog. Comp Biochem Physiol A 1990; 96:357-71.

29. Donner KO, Reuter T. Visual pigments and photoreceptor function. In: Llinás R, Precht W, eds. Frog Neurobiology. Berlin: Springer-Verlag, 1976:251-277.

30. Ingle DJ. The visual system. In: Lofts B, ed. Physiology of the Amphibians. New York: Academic Press, 1976:421-441.

31. Ingle DJ. Dishabituation of tectal neurons by pretectal lesions in the frog. Science 1973; 180:422-424.

32. Ingle DJ. Behavioral correlates of central visual function in anurans. In: Llinás R, Precht W, eds. Frog Neurobiology. Berlin: Springer-Verlag, 1976:435-451.

33. Neary TJ, Northcutt RG. Nuclear organization of the bullfrog diencephalon. J Comp Neurol 1983; 213:262-278.

34. Marin O, Gonzalez A, Smeets WJAJ. Anatomical substrate of amphibian basal ganglia involvement in visuomotor behavior. Eur J Neurosci 1997; 9:2100-2109.

35. Patton P, Grobstein P. The effects of telencephalic lesions on visually mediated prey orienting behavior in the leopard frog (Rana pipiens): II. The effects of limited lesions to the telencephalon. Brain Behav Evol 1998; 51:144-161.

36. Ewert JP, Dinges AW, Finkenstädt T. Species-universal stimulus responses, modified through conditioning, reappear after telencephalic lesions in toads. Naturwissenschaften 1994; 81:317-320.

37. Matsumoto N, Schwippert WW, Beneke TW et al. Forebrain-mediated control of visually guided precatching in toads: Investigation of striato-pretectal connections with intracellular recording/labeling methods. Behav Proc 1991; 25:27-40.
38. Muzio RN, Segura ET, Papini MR. Effects of lesions in the medial pallium on instrumental learning in the toad (Bufo arenarum). Physiol Behav 1993; 54:185-188.
39. Muzio RN, Segura ET, Papini MR. Learning under partial reinforcement in the toad (Bufo arenarum): Effects of lesions in the medial pallium. Behav Neural Biol 1994; 61:36-46.
40. Ewert JP, Finkenstädt T. Modulation of tectal functions by prosencephalic loops in amphibians. Behav Brain Sci 1987; 10:122-123.
41. Bisazza A, De Santi A, Bonso S et al. Frogs and toads in front of a mirror: Lateralisation of response to social stimuli in tadpoles of five anuran species. Behav Brain Res 2002; 134:417-424.
42. Vallortigara G. Comparative neuropsychology of the dual brain: A stroll through animals' left and right perceptual worlds. Brain Lang 2000; 73:189-219.
43. Robins A, Lippolis G, Bisazza A et al. Lateralized agonistic responses and hindlimb use in toads. Anim Behav 1998; 56:875-881.
44. Vallortigara G, Rogers LJ, Bisazza A et al. Complementary right and left hemifield use for predatory and agonistic behavior in toads. Neuroreport 1998; 9:3341-3344.
45. Deckel AW. Laterality of aggressive responses in Anolis. J Exp Zool 1995; 272:194-200.
46. Deckel AW, Jevitts E. Left vs. right-hemisphere regulation of aggressive behaviors in Anolis carolinensis - effects of eye-patching and fluoxetine administration. J Exp Zool 1997; 278:9-21.
47. Hews DK, Worthington RA. Fighting from the right side of the brain: Left visual preference during aggression in free-ranging male tree lizards (Urosaurus ornatus). Brain Behav Evol 2001; 58:356-361.
48. Hews DK, Castellano M, Hara E. Aggression in females is also lateralized: Left eye bias during aggressive courtship rejection in lizards. Anim Behav 2004; 68:1201-1207.
49. Rogers LJ, Zappia JV, Bullock SP. Testosterone and eye-brain asymmetry for copulation in chickens. Experientia 1985; 41:1447-1449.
50. Rogers LJ. Development of brain lateralization. In: Andrew RJ, ed. Neural and Behavioral Plasticity: The Use of the Chick as a Model. Oxford: Oxford University Press, 1991:507-535.
51. Vallortigara G, Cozzutti C, Tommasi L et al. How birds use their eyes: Opposite left-right specialization for the lateral and frontal visual hemifield in the domestic chick. Curr Biol 2001; 11:29-33.
52. Casperd LM, Dunbar RIM. Asymmetries in the visual processing of emotional cues during agonistic interactions by gelada baboons. Behav Proc 1996; 37:57-65.
53. Lippolis G, Bisazza A, Rogers LJ et al. Lateralization of predator avoidance in three species of toads. Laterality 2002; 7:163-183.
54. Denenberg VH. Hemispheric laterality in animals and the effects of early experience. Behav Brain Sci 1981; 4:1-49.
55. Rogers LJ. Lateralised brain function in anurans: Comparison to lateralisation in other vertebrates. Laterality 2002; 7:219-239.
56. Gosner KL. A simplified table for staging anuran embryos and larvae with notes on identification. Herpetologica 1960; 16:183-190.
57. Heuts BA. Lateralization of trunk muscle volume, and lateralization of swimming turns of fish responding to external stimuli. Behav Proc 1999; 47:113-124.
58. Wassersug RJ, Yamashita M. Assessing and interpreting lateralised behaviors in anuran larvae. Laterality 2002; 7:241-260.
59. Ewert J-P, Ingle DJ. Excitatory effects following habituation of prey-catching activity in frogs and toads. J Comp Physiol Psychol 1971; 77:369-374.
60. Burghagen H, Ewert J-P. Question of "head preference" in response to worm-like dummies during prey-capture of toads, Bufo bufo. Behav Proc 1982; 7:295-306.
61. Robins A, Rogers LJ. Lateralized prey catching responses in the toad (Bufo marinus): Analysis of complex visual stimuli. Anim Behav 2004; 68:767-775.
62. Ewert J-P, Traud R. Releasing stimuli for antipredator behavior in the common toad Bufo bufo (L.). Behavior 1979; 68:170-180.
63. Ewert J-P. Neuronal basis of configurational prey selection in the common toad. In: Ingle DJ, Goodale MA, Mansfield RJW, eds. Analysis of Visual Behavior. Cambridge, Massachusetts: MIT Press, 1982:7-45.
64. Wachowitz S, Ewert JP. A key by which the toad's visual system gets access to the domain of prey. Physiol Behav 1996; 60:877-887.
65. Rogers LJ, Anson JM. Lateralisation of function in the chicken forebrain. Pharmacol Biochem Behav 1979; 10:679-686.

66. Andrew RJ, Mench J, Rainey C. Right-left asymmetry of response to visual stimuli in the domestic chick. In: Ingle DJ, Goodale MA, Mansfield RJW, eds. Analysis of Visual Behavior. Cambridge, Massachusetts: MIT Press, 1982:197-209.
67. Deng C, Rogers LJ. Differential contributions of the two visual pathways to functional lateralization in chicks. Behav Brain Res 1997; 87:173-182.
68. Güntürkün O. Lateralization of visually guided behavior in pigeons. Physiol Behav 1985; 34:575-577.
69. Güntürkün O, Kesh S. Visual lateralization during feeding in pigeons. Behav Neurosci 1987; 101:433-435.
70. Alonso Y. Lateralization of visual guided behavior during feeding in zebra finches (Taeniopygia guttata). Behav Proc 1998; 43:257-263.
71. McKenzie R, Andrew RJ, Jones RB. Lateralization in chicks and hens: New evidence for control of response by the right eye system. Neuropsychologia 1998; 36:51-58.
72. Dharmaretnam M, Andrew RJ. Age- and stimulus-specific use of right and left eyes by the domestic chick. Anim Behav 1994; 48:1395-1406.
73. Andrew RJ. Left and right hemisphere memory traces: Their formation and fate. Evidence from the events during memory formation in the chick. Laterality 1997; 2:179-198.
74. Rashid N, Andrew RJ. Right hemisphere advantage for topographical orientation in the domestic chick. Neuropyschologia 1989; 27:937-948.
75. Traub B, Elepfandt A. Sensory neglect in a frog: Evidence for early evolution of attentional processes in vertebrates. Brain Res 1990; 530:105-7.
76. Ewert JP. Concepts in vertebrate neuroethology. Anim Behav 1985; 33:1-29.
77. Güntürkün O. Visual lateralization in birds: From neurotrophins to cognition? Eur J Morphol 1997; 35:290-302.
78. Güntürkün O. Morphological asymmetries of the tectum opticum in the pigeon. Exp Brain Res 1997; 116:561-566.
79. Deng C, Rogers LJ. Light experience and lateralization of the two visual pathways in the chick. Behav Brain Res 1999; 98:277-287.

SECTION III
Vertebrate Studies
of Physiological Asymmetries—
Perspectives from the West and the East

CHAPTER 9

The Evolution of Behavioral and Brain Asymmetries:
Bridging Together Neuropsychology and Evolutionary Biology

Giorgio Vallortigara*

Abstract

The evidence for brain and behavioral lateralization in human and nonhuman species is reviewed and discussed in an evolutionary perspective. It is stressed that current theories of the evolution of lateralization and of its alleged biological advantages fail to acknowledge the riddle of the alignment of the direction of asymmetries at the population level. The different specialization of the left and right side of the brain has been supposed to increase brain efficiency. However, lateral preferences in behavior that arise as a consequence of brain asymmetries usually occur at the population level, with most individuals showing similar direction of bias. Individual brain efficiency does not require the alignment of lateralization in the population. Alignment of the direction of behavioral asymmetries in a population can arise when individually asymmetrical organisms must coordinate their behavior with that of other asymmetrical organisms. Thus, population lateralization may have evolved under social selection pressures as an example of an "evolutionarily stable strategy".

Introduction

Three very important characteristics of our species, language, right-handedness and tool use, have been traditionally associated with a single and (allegedly) unique characteristic of the human brain, namely hemispheric specialization. The phenomenon of hemispheric specialization (or brain lateralization or asymmetry) refers to the different functional specialization of the left and right side of the brain. For instance, in most (right-handed) individuals of our species the brain mechanisms for language production are located in the left hemisphere.[1]

Although research on human cerebral lateralization has a long tradition, until very recently we knew very little on lateralization as a biological phenomenon. This is because of the erroneous assumption that lateralization is a uniquely human attribute. Research in the last years has shown that this is not true. Lateralization is in fact widespread among vertebrates, and it is not at all unique of the human brain (see for recent reviews refs. 2-4). Animal models can now be used to gain insights into the neuronal processes governing lateralized functions. Moreover, with the support of animal system models we can now try to answer questions concerning the function, other than the structure, of cerebral lateralization.

*Giorgio Vallortigara—Department of Psychology and B.R.A.I.N. Centre for Neuroscience, University of Trieste, Via S. Anastasio 12 34123 Trieste, Italy. Email: vallorti@univ.trieste.it

Behavioral and Morphological Asymmetries in Vertebrates, edited by Yegor B. Malashichev and A. Wallace Deckel. ©2006 Landes Bioscience.

The great ethologist Niko Tinbergen[5] identified four types of questions that should be asked about behavior: (1) What are the mechanisms that cause it? (2) How does it develop? (3) How did it evolve? (4) What is its function, namely its survival value?

In this paper I'll try to apply this ethological style of analysis to the problem of cerebral lateralization. I will concentrate in particular on the work carried out in my laboratory in the last 15 years.

Proximal Mechanisms: Lateralized Cognition and Neural Events

Let us start with causal mechanisms. Here I would like to focus on what sort of quite peculiar advantages we can gain from the use of some animal models, in particular from the use of the avian brain.

In animals with laterally placed eyes, such as most species of birds, there is a virtually complete decussation at the optic chiasm. In the optic nerves less than 0.1% of the fibres proceed to the ipsilateral side.[6] Since only a limited number of axons recross via the mesencephalic and thalamic commissures, the avian visual system is remarkably crossed. This means that information entering each eye is largely, though not completely (see refs. 7,8), processed by the contralateral side of the brain. Thus, by simply temporarily occluding one eye we can obtain some insights on lateralized functions of the avian brain. Let us consider some examples of the types of data we can obtain with this very simple technique. I'll concentrate on spatial tasks and the domestic chick as a model, but other functions and species will be considered later on.

Figure 1, depicts a delayed-response task developed in my laboratory (see refs. 9,10). The chick is confined within a transparent partition from where it can observe an imprinting object moving out-of-sight behind a screen. After a while the chick is allowed to rejoin the imprinting object. At test, the chick is presented with two screens. In each trial the object is made to disappear behind either one of the screens at random. After a delay (30 sec), during which an opaque cover is placed in front of the transparent cage to prevent viewing of the screens, the chick is allowed to search for the disappeared object. In one version of the task, we used two

Figure 1. The delayed-response task developed in the laboratory of the author. The imprinting object is made to disappear behind either one of two opaque screens, meanwhile the chick is confined within a transparent enclosure. After a 30 sec delay, the chick is then released and allowed to approach the screens (chicks can manage even with delays up to 60 sec and more; see refs. 9,10,103). During the delay (when an opaque partition prevented the chicks having any sight of the screens) the right-left position of the two screens was swapped rapidly. When released, the chick was thus faced with the problem of choosing either the correct screen in the wrong position or the incorrect screen in the correct position.

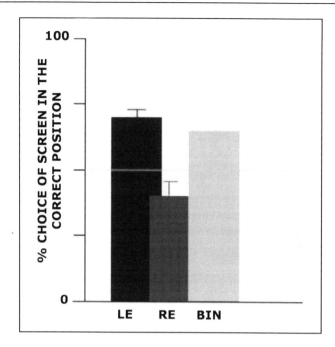

Figure 2. Results of an experiment with the delayed-response task when position- and object-specific cues associated with the screens were contrasted. Chicks using their left eye searched the imprinting object behind the screen in the correct position and the wrong pattern. Chicks using their right eye did the reverse. Chicks using both eyes behaved like left-eyed chicks.

screens with different colour and, during the delay, we interchanged their spatial position. In this way after opening of the cage the chicks were faced with the problem of choosing the screen in the correct position but with the wrong colour, or the screen with the correct colour but in the wrong position. The results (Fig. 2) showed that left-eyed (LE) chicks tended to search behind the screen in the correct position, and right-eyed (RE) chicks behind the screen of the correct colour (see ref. 2). So the right hemisphere attends to the position of the screen and the left hemisphere to the colour of the screen. Note, also, that the behavior of chicks using both eyes (BIN) is identical to that of LE chicks. This means that in the normal, binocular condition of vision there is one hemisphere—the right in this case—that takes charge of control of behavior.

Another, even more striking example of dissociation between the use of position-specific and object-specific cues is shown in Figure 3 (see ref. 2). Chicks were trained to find food hidden below sawdust on the floor by ground scratching in the centre of a closed uniform arena: the centre was indicated by a conspicuous landmark (a red stick). After learning, the landmark was dislocated in a corner and chicks were tested binocularly or with only one eye in use. A striking asymmetry appeared (Fig. 3): binocular chicks and chicks using only their left eye searched at the centre (ignoring the landmark), whereas chicks using only their right eye searched at the corner with the landmark (ignoring spatial information).

We then restricted our search for lateralized mechanisms, looking for specific areas in the left and right hemispheres, which were differentially involved in these tasks. The hippocampus is very important for memory in humans and there is evidence for a predominant role of the right hippocampus in spatial navigation.[11] We performed experiments in chicks whose left or right hippocampus was lesioned.[12] Control, sham-operated, chicks behaved as binocular chicks, searching in the centre and ignoring the landmark. Bilateral- and right-hippocampal lesioned

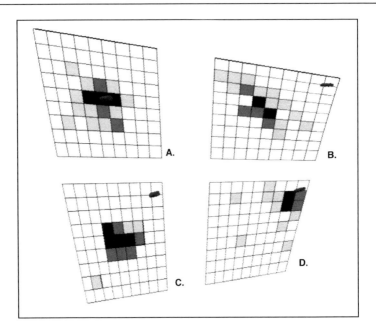

Figure 3. Chicks were trained binocularly (A) to ground scratch in the center of an enclosure to find food hidden below sawdust. The position of the food was indicated by a landmark (a red stick) and by geometric information provided by distances from the walls of the enclosure. Chicks were then tested following the displacement of the landmark to a corner. The dark areas represent the area of search (redrawn from data in refs. 2,14). After the landmark displacement, left-eyed (C) and binocular (B) chicks continued to search in the centre, whereas right-eyed chicks (D) searched in the corner, near the landmark.

chicks, in contrast, searched near the landmark, whereas left-hippocampal lesioned chicks were midway between choosing the centre and choosing the landmark.

An even more striking asymmetry appeared following the simple removal of the landmark. In this case only the chicks with an intact right hippocampus were able to orient using the residual information provided by the distances from the walls of the enclosure. Apparently, in the absence of the right hippocampal formation chicks could rely only on local information, the landmark, and were completely impaired in the use of large-scale, geometric, spatial information (and see also ref. 13).

It should be noted, however, that not all spatial orientation tasks in animals are right-hemisphere biased. We found that different aspects of spatial cognition are represented in the left and right hemisphere (see refs. 14-16). We trained chicks to find food by ground scratching on the floor of a closed enclosure. This time, however, there was no landmark to indicate the position of food. Only the distance from the walls allowed the chick to identify the food position in the centre. Chicks managed quite well in this task and generalized to enclosures of roughly the same size but different shape. One interesting result emerged when the chicks were placed in larger enclosures of the same shape. In this case chicks exhibited two strategies for searching behavior: one based on absolute distances and one based on relative distances.[14,17] We wondered whether this could be the result of different strategies exhibited by the left and the right hemisphere. In Figure 4, the amount of searching behavior is depicted as a function of distance from the centre. Results for training in a small arena and then testing in a larger arena are shown. Double peaks of search in binocular chicks are found when they are tested in the larger arena. Left- and right-eyed chicks are strikingly different: chicks using their left hemisphere searched only at the absolute distance, chicks using their right hemisphere searched only at the relative distance.

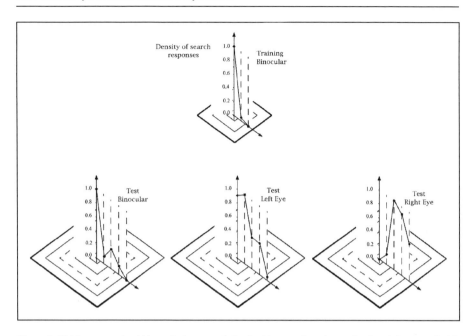

Figure 4. Chicks were trained binocularly to search for food in a small enclosure (top) and then tested with only one eye in use in a larger enclosure. During test binocular chicks showed one peak of search behavior located in the centre of the large enclosure and one peak located at a distance from the walls corresponding to the previously learnt distance from the centre in the training (small) enclosure. Left eyed chicks showed mainly a peak located in the centre of the large enclosure. Right-eyed chicks, in contrast, showed only one peak located at the distance from the walls corresponding to the previously learnt distance from the centre in the training (small) enclosure.

How can these phenomena be explained at the neuronal level? Recent work with single cell recordings carried out by Onur Güntürkün and colleagues (Bochum University, Germany) provide some insights.[18] By recording from the left or the right n. rotundus while using a standardized visual stimulation paradigm of the ipsi- or the contralateral eye, these researchers distinguished between left-right differences that emerged bottom-up from the retino-tecto-rotundal system from those that were derived top-down from the forebrain (see ref. 19). Left-right differences within the bottom-up system were due to variations in the latency and the tonic spike duration of rotundal neurons after stimulation of the contralateral eye. Visual signals arrived on the average 18% faster in the right thalamus, but cellular activation lasted 27% longer in the left rotundus. The authors suggested that these lateralized effects may underlie the fact that pigeons are faster with the left eye (right hemisphere) in simple visual reaction paradigms, but are superior with the right eye (left hemisphere) in pattern learning and discrimination. While the asymmetries within the bottom-up system were a matter of degree, those of the top-down cells displayed an all-or-none organization. All thalamic cells activated by descending forebrain systems were under the control of the left hemisphere. Thus, although visual input reaches both hemispheres, the modulation of the diencephalic relay of the tectofugal system seems to be under the executive control of only the left hemisphere.

An intriguing consequence of these data is that if descending forebrain signals arrive within the rotundus only from the left hemisphere, they should produce response patterns with diverse combinations of bottom-up and top-down influences depending on the thalamic side. Within the left thalamus most bottom-up and all of the top-down effects are communicated by the right eye system. This is different for the right rotundus, where bottom-up input derives from left eye stimulations while all of the top-down effects originate from the right eye input.

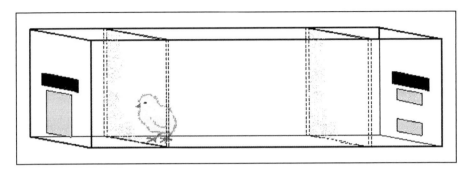

Figure 5. The apparatus employed for the 6-min free-choice test which chicks underwent on day 3 in order to investigate hemispheric differences in perception of partly occluded objects (see Regolin et al[23] for details). The most crucial condition is schematically represented: monocular chicks that had been imprinted (binocularly) onto the occluded square could now choose between an amputated version of it (physically identical to the visible parts of the imprinting object itself) and a complete, not occluded square.

Thus, bilateral integration predominates at right thalamic level but not at left. This electro-physiological pattern could explain some of the well-known asymmetries on spatial orientation and attentional control. In most species studied, the right hemisphere dominates visuospatial orientation and attention, a cognitive feature that generally requires the integration of information from widespread areas of the visual field.

These electrophysiological data fit very well with some recent cognitive evidence coming from studies on the abilities of young chicks to mentally complete partially occluded objects. Evidence collected both in newly-hatched chicks[20] and adult hens[21,22] suggests how this bird species is capable of perceiving as a whole objects that are partly concealed by occluders ("amodal completion"). We recently investigated hemispheric differences in amodal completion by testing newborn chicks with one eye temporarily patched.[23] Separate groups of newly hatched chicks were imprinted binocularly (i) on a red cardboard square partly occluded by a superim-posed black bar, (ii) on a whole version of the red square, or (iii) on an amputated version of it (see Fig. 5). At test, in monocular conditions, each chick was presented with a free choice between a complete and an amputated square. In the crucial condition (i) chicks tested with only their left eye in use picked the completed square (like binocular chicks would do); chicks with only their right eye in use, in contrast, tended to choose the amputated square.[23] Similar results were obtained with chicks imprinted binocularly onto a red cardboard cross (either occluded or amputated in its central part). When asked to choose between a complete and an amputated cross, left-eyed and binocular chicks chose the complete cross, whereas right-eyed chicks showed a trend for choosing the amputated cross (see ref. 23).

The above results suggest that the right hemisphere controls the process of amodal comple-tion of partly occluded objects. Interestingly, even in humans the right hemisphere plays a more important role in amodal completion (see ref. 24). Note that in order to amodally complete an object, the spatial relationships between the parts of a visual scene must be taken into account, a task in which the right hemisphere excels.[2,14,25,26] It seems, therefore, that the right hemisphere is more specialized at detecting the global structure of visual objects, while the left hemisphere might be more inclined at detecting local features. Moreover, the right hemisphere may be "in charge" of control of behavior in these tasks as evinced by the fact that binocular chicks behaved similarly to left-eyed chicks.

We hypothesize, however, that these hemispheric differences are mostly a matter of degree rather than of kind. In the natural condition, when birds can use freely both hemifields, the two strategies of holistic and analytic visual analysis should reciprocally support each other rather than compete. Thus hemispheric differences may modulate, possibly by attentional mechanisms, the type of analysis to be carried out on visual stimuli. Evidence suggests that birds can activate the hemisphere most appropriate to particular conditions and stimuli by using lateral fixation with the contralateral eye.[3,27,28] Similarly, lateralized mechanisms somewhat akin to those available in birds have been described in the human neuropsychological literature, for instance in the form of lateralized direction of gaze or voluntary eye movements to the left or to the right associated with the type of hemispheric strategies to bring into play (e.g., ref. 29).

The Development of Brain Lateralization

The existence of lateralization in the behavior of newly-hatched chicks suggests that it is an inborn characteristic of the brain. However, there is also striking evidence for a role of experience in the development of lateralized behavior. In the last few days before hatching, the embryo of most species of birds turns in the egg so that it occludes its left eye. Only the right eye is exposed to light entering through the eggshell and membranes. Rogers and colleagues (reviews in refs. 8,30-32) have shown that exposure to light for at least 2 hours on day 19/20 of incubation causes an alteration in a variety of lateralized behaviors. For example, on the pebble floor task in which chicks have to discriminate grains of food from pebble scattered on the floor, right-eyed chicks coming from eggs exposed to light learn the task. Left-eyed chicks fail to learn this task. The asymmetry disappears in chicks coming from eggs maintained in darkness during the critical period. It is also possible to reverse the direction of the asymmetry, by turning the head of the embryo and exposing it to light on the left side.[33] Similar results have been obtained in pigeons.[34,35]

In experiments in which fluorescent tracers were injected into the visual Wulst in the telencephalon, in order to map ipsi- and contralateral backward projections in the thalamus, asymmetric light exposure in the chick appeared to generate anatomical asymmetries in the thalamofugal pathway. The exposure of the right eye to light was associated with an increased number of visual projections from the left side of the thalamus (which receives inputs from the right eye) to the right Wulst. Conversely, no changes in the number of projections from the right side of the thalamus to the left visual Wulst were found.[8]

Intriguingly, in the pigeon a similar asymmetry has been found, but in the tectofugal rather than in the thalamofugal pathway. The exposure of the right eye to light leads to an increased number of visual projections from the right tectum to the contralateral rotundus (see ref. 36).

There are, however, forms of population-level lateralization in chicks that do not depend (or that do not depend entirely) on light exposure of the embryo. These include imprinting and social recognition,[37-39] response to olfactory versus visual cues,[40] lateralization of auditory responses[41] and lateralization of spatial cognition.[42]

A form of lateralization that occurs in chicks coming from eggs maintained in the dark is associated with unihemispheric sleep. Chicks (and other species of birds) during sleep show brief and transient periods in which one eye is opened while the other remains closed.[43] Electrophysiological recordings revealed that the hemisphere contralateral to the open eye shows an EEG with fast waves typical of wakefulness. Conversely, the hemisphere contralateral to the closed eye shows an EEG typical of slow sleep. Recently we explored monocular sleep in chicks coming from eggs exposed to light or maintained in the dark.[44] As shown in Figure 6, light-exposed chicks show right-eye opening during sleep in the first two days after hatching, and then turn to left-eye opening. Dark-incubated chicks, in contrast, show consistent left-eye opening preferences during sleep from soon after hatching throughout day 5. Clearly this bias cannot be due to light stimulation in the embryo as it seems to be already present as an inborn condition.

Direct evidence that lateralization may depend on genetic mechanisms has been obtained in my laboratory using fish. We used the teleost fish, *G. falcatus*, in a detour task. The fish were free to move in a tank that faced, on both sides, a barrier of vertical sticks behind which was a dummy predator. Direction of turning reflected preferences for the use of the lateral (monocular) field of the left and right eye (see ref. 45). Strongly left- or right- eye preferent fish were mated

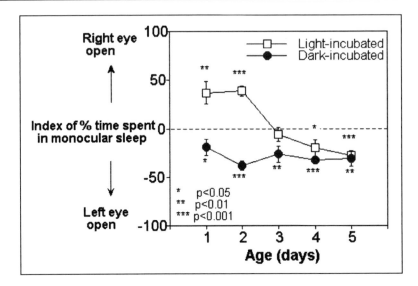

Figure 6. Time spent with the left or right eye opened during unihemispheric, monocular sleep in chicks coming from eggs incubated in the light or in the dark.

together. The results obtained with the progeny showed that individual differences in lateralization can be inherited, both in strength and direction.[46] Our estimate of heritability in *G. falcatus* was quite high (about 0.56) and accounted for more than 50% of the variance in eye preference in the offspring.[46] These findings suggest that the turning bias in *G. falcatus* has a strong genetic component.

More recently using this species of fish we developed two lines of animals that turn right (RD) in the detour test, two lines of fish that turn left (LD) and one unselected control line.[45] Fish from the LD and RD lines were then compared using several different tests of lateralization. To date, comparison has been completed for five tests, two of which tested motor lateralization and the remaining three visual lateralization. The results of the experiments showed that the two lines segregate very clearly in all tests.[47] Thus, behavioral asymmetries in the detour test are predictive of lateralization in other types of behavioral tests. Moreover, these results show that RD and LD fish have a similar but left–right reversed pattern of subdivision of cognitive/behavioral functions, which is suggestive of a similarly left–right reversed (mirror image) brain organization.

Genetic analyses of lateralization at the molecular level in fish promises to be a very rewarding area of research in the near future. Several groups are now working on this area using the zebrafish (*Danio rerio*), a system model widely used in developmental biology (e.g., refs. 48-50). As in many lower vertebrates, the zebrafish epithalamus shows multiple left-right differences. These include a parapineal on the left of the pineal, expanded dense neuropil in the left habenular nucleus, and asymmetric patterns of habenular gene expression.[51] Through genetic manipulations, future work will be able to randomize directional asymmetry of the larval brain, and rear "left-biased" and "right-biased" fish separately, as well as to visualize laterality of the brain in live transgenic animals.

The Evolution of Brain Lateralization

Now we can move from development to evolution. Here we will consider the questions, (a) what species are lateralized?, and (b) how did lateralization evolve in the different phylogenetic lineages?

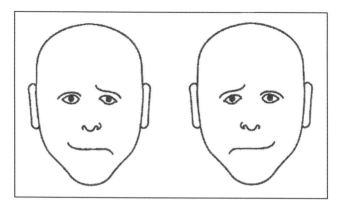

Figure 7. Look at the nose of each face and say which individual looks happier. Usually, for most people the face on the left, with his happy half on the left, looks happier.

Evidence that nonhuman mammals also have lateralized brain is widespread. One nice example concerns emotional expression in the face. Before reading any further, turn to Figure 7, and look directly at the nose of each face. Decide which individual face looks happier. Most right-handed people say that the face with the smile placed on the left looks happier. This is due to two phenomena, i.e., that greater attention is paid to the left side of the face and also that the right hemisphere is better at recognizing emotions than the left hemisphere. The converse is also true, in that emotions are expressed more strongly in the left side of the face.

Asymmetrical processing of emotion also occurs in monkeys. Marc Hauser at Harvard did a frame-by-frame analysis of the fear grimace of rhesus monkeys produced in the context of submission. He found a temporal asymmetry in the facial expression of emotion. Specifically, the left side of the face showed the fear grimace more quickly than the right side of the face (ref. 52 and see also ref. 53).

The right hemisphere in humans is dominant in recognizing faces (see ref. 54). Prosopagnosia, the inability to recognize familiar faces, may occur following right hemisphere damage.[55] The inferotemporal cortex is involved in these abilities[56] and the same seems to be true for sheep.[57] Keith Kendrick at Babraham Institute in Cambridge found cells in the inferotemporal cortex of sheep that respond selectively to the view of a face, or even to special features of a face, like the horn. Recently, Kendrick also showed that cells in the right temporal cortex encoding face categories or individual faces respond much faster and with greater synchrony than cells in the left hemisphere (review in ref. 58).

The fact that mammals, birds and fish all show brain lateralization strongly suggests that it has been inherited in all these taxonomic groups from some ancient common vertebrate ancestor. The hypothesis of independent evolution (homoplasy rather than homology) seems unlikely. Indeed, lateralization has been observed even in amphibians and (as we shall see later) in reptiles too. For instance, we found that three different species of toads, *B. marinus*, *B. viridis* and *B. bufo* show preferential use of the right forepaw to remove a piece of paper stuck on their mouth.[59] Lateralization in amphibians has been more recently confirmed and further investigated in several other laboratories (see for reviews refs. 60,61).

A Digression: Lateralization in the Wild

Until very recently, research on brain lateralization has been confined within the laboratory of experimental psychologists and neuropsychologists and largely centered on the human species. Despite the impressive body of evidence showing the existence of lateralization in nonhuman species, extending from fish to primates, the subject has remained largely outside the realm of biology. In the last few years, however, evidence for lateral biases affecting everyday behavior in the natural environment of a variety of species has been published. This evidence forces a rethinking of some of the basic issues related to the evolution of lateralization. Below, I briefly review some recent work that has been done on this topic.

It has been shown that toads attack conspecifics to the left and strike preferentially at prey on the right.[62] In this work a preferred prey was mechanically moved in a horizontal plane around the toad. The prey was delivered to either the right or left monocular visual field depending on the direction of rotation (see ref. 62). When the prey moved clockwise, and thus entered first the left and then the binocular field of vision, almost all of the tongue-strikes occurred in the right half of the binocular field. When the prey moved anticlockwise, and thus entered first the right and then the binocular field of vision, a more symmetrical distribution of strikes in the left and right halves of the binocular fields occurred. Thus, it seemed necessary for prey to enter the right half of the binocular visual field in order to evoke predatory behavior. Functionally, it was as if the toads' left hemifield showed a form of a stimulus-specific visual "hemineglect" for predatory behavior. (Similar forms of "pseudoneglect" have been demonstrated more recently in birds; ref. 63).

The same toads were also tested for agonistic behaviors in the form of tongue-strikes at competitors during feeding.[62,64] Toads showed a population bias to strike with the tongue at conspecifics when these were occupying their left visual field. Thus toads are more likely to attack a prey to their right side (and ignore them to their left side) and to attack a conspecific to their left side (and ignore them to their right side).

Similar results have been obtained in a variety of other species. For instance in the domestic hen,[65] in a lizard, *Anolis carolinensis*, the American chameleon[66,67] and in the gelada baboon.[68]

To test the response to predators, we presented toads with a simulated predator entering suddenly into their left or right monocular or binocular visual fields.[69] We found that toads were more likely to react, most often by jumping away, when a simulated predator was in their left monocular field than when in their right monocular field.

More recently, we documented complementary eye use in naturalistic settings in a bird species, the Black winged stilt (*Himantopus himantopus*) during predatory pecking, courtship and mating behavior.[70] We found that black-winged stilts have a population-level preference for using their right monocular visual field before predatory pecking; moreover, pecks that followed right hemifield detection were more likely to be successful than pecks that followed left hemifield detection, as evidenced by the occurrence of swallowing and shaking head movements after pecking. In contrast, shaking behavior, exhibited as part of courtship displays, and copulatory attempts by males were more likely to occur when females were seen with the left monocular visual field.

As a final example particularly relevant to the hypothesis I am going to put forward, it is commonly observed that some species of fish leave their shoal in pairs in order to approach and inspect a potential predator.[45] The risk of being preyed upon is shared if both fish simultaneously inspect the predator, but not if one of the fish remains at a distance. This situation can be simulated in the laboratory with the aid of a mirror (see ref. 71). We found that fish (*Gambusia holbrooki*) were more likely to approach a predator for inspection when the mirror was on their left side than when the mirror was on their right side.[72] Such a preference for a fish to position itself so that the image of a conspecific is on its left side has been reported, even in the absence of predators, in eight different species of teleosts.[73-75] Similarly, there is a preference for several species of teleosts to position themselves, even in the absence of other conspecifics, so that the image of a predator during inspection is on their right side.[76,77]

Overall, this evidence clearly shows that asymmetries in overt behavior in everyday life are the norm rather than the exception among vertebrates. This brings us to the last question, i.e., what are these asymmetries for; what is their function?

What Is Lateralization For? Costs and Benefits of an Asymmetrical Brain

The physical world is indifferent to left and right side of interpersonal space. A deficit on one side both would leave an animal vulnerable to attack from that side and leave it unable to attack prey or competitors appearing on that side. Thus there appear to be striking disadvantages to possessing a perceptual system that is asymmetrical to any substantial degree. However, as we have seen, ethological observations tell us that such asymmetries are not at all rare, but quite ubiquitous in animal behavior.

A possible explanation for this state of affairs would be to argue that the benefits associated with possession of an asymmetric brain could counteract the ecological costs associated with lateral biases in overt behavior.[78] A crucial benefit that lateralization may offer is to increase neural capacity, because specializing one hemisphere for a particular function leaves the other hemisphere free to perform other (additional) functions. This would allow brain evolution to avoid useless duplication of functions in the two hemispheres, thus sparing neural tissue. Another advantage of lateralization is that dominance by one hemisphere (or in general by one side of the brain) is likely to be a convenient way of preventing the simultaneous initiation of incompatible responses in organisms with laterally placed eyes.[79-82] More generally, lateralization could be one way of increasing the brain's capacity to carry out simultaneous processing, by enabling separate and parallel processing to take place in the two hemispheres. For example, Lesley Rogers, Paolo Zucca and myself recently tested chicks on a dual task, one involving the left hemisphere in control of pecking responses and the other involving the right hemisphere in monitoring overhead to detect a model predator.[83] Chicks exposed to light before hatching were compared to those incubated in the dark, since the light exposure aligns and strengthens visual lateralization on these tasks. Rogers[84] showed that strongly lateralized (light-exposed) chicks detected the model predator sooner with the left eye (i.e., when the right hemisphere was attending to the stimulus) than did the weakly lateralized (dark-incubated) ones. We confirmed this result by scoring not only the response to the model predator but also the chick's ability to learn to peck at grain versus pebbles.[83] Strongly lateralized chicks avoided pecking at pebbles far better than weakly lateralized chicks and they were also more responsive to the model predator. As a control, we tested the weakly lateralized chicks on the pebble-grain test without presenting the model predator and found that they had less difficulty in learning to discriminate grain from pebbles. Hence, the weakly lateralized chicks had their greatest difficulties when they attended to the two separate tasks simultaneously.

There is a problem, however, in viewing lateralization as an advantage in the computational abilities of the brain. The asymmetries we have described so far are "population-" or "species-level" asymmetries, i.e., asymmetries showing a similar direction in more than 50% of the population (see also ref. 85). In biological terms these are "directional" asymmetries, quite distinct from asymmetries occurring at the "individual-level", i.e., asymmetries showing an equiprobable distribution within a population, with 50% of the individuals favoring the left and 50% favoring the right (see ref. 85).

Although the hypothesis of a computational advantage may explain individual lateralization, it does not, in itself, explain the alignment in the direction of lateralization at the population level. In fact, individual brain efficiency is unrelated to how other individuals are lateralized. Why, therefore, do most animals (usually 65-90%) possess a left eye (or hemifield) better suited than the right for vigilance against predation? Would it not be simpler for brain lateralization to be present in individuals without any specification of its direction (i.e., with a 50:50 distribution of the left- and right-forms in the population)?

One can argue that the alignment of the direction of lateralization at the population level is a mere by-product of genetic expression. However, this cannot be so because we know that selection for the strength (without direction) of lateralization is possible. This has been shown for instance by Collins[86] with mice. A further puzzle is that such an alignment may even be disadvantageous, as it makes individual behavior more predictable to other organisms. If most of the fish of a population turn leftward when encountering a predator (see e.g., ref. 87), the predator can learn about the bias of the prey and exploit it during prey catching. The same would not hold if prey were only individual-level lateralized. Thus, there should have been important selective pressures to maintain directional asymmetry in spite of its potential disadvantage.

It is also noteworthy that there is clear evidence that individual asymmetries alone can be of advantage to the fitness of organisms. Consider foraging for termites by wild chimpanzees. Both hands are used by chimpanzees in termite fishing, one to hold the twig used as a probe and the other to act as a stabilizer across which the twig covered in termites is rubbed when the chimpanzee eats them. There is no evidence for a population-level bias in wild-chimpanzees studied at Gombe in this task; nonetheless there is individual lateralization, with some individuals preferentially using the same hand to probe and the other to stabilize the twig (lateralized), whereas others vary which hand is used for either purpose (ambidextrous). McGrew and Marchant[88] studied the efficiency of termite fishing by the chimpanzees and found that individually lateralized chimpanzees, irrespective of the direction of their lateralization, gathered more prey for a given amount of effort than did ambidextrous chimpanzees. Thus, individual lateralization clearly suffices to confer a fitness advantage without any need for an alignment of lateralization at the population level.

If lateralization at the individual level suffices both logically and empirically to produce computational (and thus fitness) advantages, why then do we observe population-level asymmetries for a wide range of vertebrates on many tasks? Here I suggest that a simple concept may explain this, i.e., that sometimes what is better for an (asymmetrical) individual to do depends on what the other (asymmetrical) individuals of the group do. In other words, I suggest that there maybe "social" constraints that force individuals to align their asymmetries with those of the other individuals of the group (see Vallortigara and Rogers[89] for a more extended discussion of the hypothesis).

The idea that lateralization may convey a social advantage was first put forwards by Lesley Rogers in 1989 when observing that the social hierarchy was more stable in groups of light-exposed (lateralized) chicks compared to groups of dark-incubated (not-lateralized) chicks (see ref. 90). However, the fact that there is more stability in a group, by itself, does not explain the alignment of lateralization in the different individuals, for any fitness benefit should be computed at the level of the individual (or, better, at the level of the gene), not at the level of the group. We recently worked out a more detailed hypothesis, i.e., that the alignment of the direction of behavioral asymmetries in a population can arise as an **evolutionarily stable strategy** (ESS, see ref. 91), when individually asymmetrical organisms must coordinate their behavior with that of other asymmetrical organisms.

With Stefano Ghirlanda I formalized a game theory model showing that population-level lateralization can indeed be evolutionarily stable.[92] The model was framed in the context of prey-predator interactions, but can be extended to other scenarios. Here I provide a synthetic version of it.

Consider prey-predator interaction. Prey lateralized in the same direction have a greater chance of keeping together as a group. On the other hand, predators may learn to anticipate prey escape movements, or to approach prey from a given direction. Let us write as $p(x)$ the probability that a prey survives an attack, given that a proportion x of its group-mates has its same lateralization. A simple yet fairly general way of writing $p(x)$ is:

$$p(x) = p_0 + cg(x) - l(x) \tag{1}$$

where p_0 is a baseline escape probability, $g(x)$ represents the benefit gained, under attack, by keeping together with a proportion x of fellow prey, and $l(x)$ represents the cost of having the same directional bias as a proportion x of other prey (both $g(x)$ and $l(x)$ are assumed positive). This cost is assumed to arise from predators having more success with the more common prey type. The parameter c allows to regulate the relative importance of $g(x)$ and $l(x)$.

If we indicate with a and $1-a$, respectively, the proportion of left- and right-type prey in the population, we can use equation (1) to write the respective escape probabilities as:

$$p(a) = p_0 + cg(a) - l(a)$$

$$p(1-a) = p_0 + cg(1-a) - l(1-a)$$

(2)

The condition for a given proportion a^* to be an evolutionary equilibrium is that the escape probabilities of left- and right-type prey be equal, that is

$$p(a^*) = p(1-a^*)$$

(3)

The equilibrium is stable if natural selection works to restore the proportion a^* whenever slight deviations occur. This means that a small increase in the proportion of left-type prey, say by an amount ε, should increase the escape probability of right-type prey, and vice-versa. In formulae:

$$p(a^* + \varepsilon) < p(1 - a^* - \varepsilon)$$

$$p(a^* - \varepsilon) > p(1 - a^* + \varepsilon)$$

(4)

These equations provide us with a simple, general framework to study the evolutionary stability of populations composed of left- and right-type prey. The existence and nature of equilibria depends, of course, on what $g(x)$ and $l(x)$ are.

We have linked $l(x)$ to the ability of predators at capturing a given prey type, as a function of this type's abundance. That is, this function should measure the performance of predators as a function of the amount of practice with a given prey type. Empirically, performance curves of this kind are often well approximated by a negatively-accelerated function:

$$l(x) = 1 - \exp(-kNx)$$

(5)

where N is group size, and larger values of the positive parameter k lead to faster increase of performance with increasing prey abundance.

As to group effects on predation risk, one relatively well-studied effect is so-called "dilution", whereby in a group of n each individual is assumed to have a probability of $1/n$ of being targeted by a predator.[93] This probability can be approximated by $1/(1+Nx)$ if a prey keeps together with a fraction x of individuals from a larger group of N. The probability of not being chosen as target is therefore,

$$g(x) = 1 - \frac{1}{1 + Nx}$$

(6)

This expression can be used in equation (1) as the benefit of group living to an individual prey, when a proportion x of prey is using its same strategy (since prey with the same strategy are assumed to be more likely to keep together).

Employing equations (5) and (6), together with the equilibrium and stability conditions (3) and (4), Ghirlanda and Vallortigara[92] analyzed numerically the existence and stability of equilibria. In Figure 8, is shown the equilibrium proportion of left-type prey as a function of the parameter c in equation (1) and for $N = 50$, $k = 0.25$.

The figure shows that for small c the only stable population consists of left- and right-type prey in equal numbers. This corresponds to situations in which lateralization-mediated effects of group living on escape probability are small (see eqn. (1)), for instance in the case of solitary prey or for lateral biases that do not influence group cohesion. This equilibrium becomes unstable for larger c (larger group effects), giving way to stable populations consisting of left- and right-type prey in unequal numbers. Since the model does not assume any intrinsic benefit of left or right lateralization, there are always two specular solutions, one with a majority of left-type prey and one with a majority of right-type prey. The intuitive content of such a situation is that the majority of prey get protection by keeping together, but pay a cost because predators are

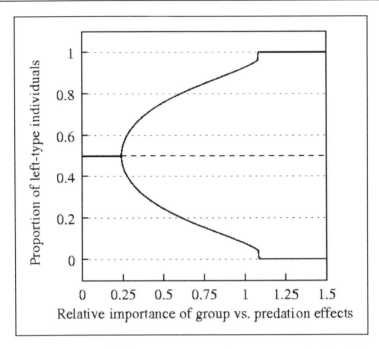

Figure 8. Equilibrium proportion of left-type prey in a group living species as a function of the parameter *c* in equation (1) (see text for details). Solid lines: stable equilibria; dashed lines: unstable equilibria. (Redrawn from data of Ghirlanda and Vallortigara[92]).

better at handling them. A minority of prey manages to enjoy the same escape probability by trading-off protection from the group with an advantage in the face of predators. Figure 8, also shows that the proportion of the majority prey type increases as *c* gets larger, until only populations composed entirely of one type of prey are stable. This corresponds to situations where the protection offered by the group is so large as to overcome any effect of differential ability in predators. For smaller values of *c*, populations composed of only one prey type are always unstable.

Thus, the model shows that populations consisting of left- and right-type individuals in unequal numbers—the most common situation among vertebrates—can be evolutionarily stable if being lateralized in one or the other direction has frequency-dependent costs and benefits.

From a mathematical point of view, therefore, the hypothesis is sound. But how can the hypothesis be tested empirically? As an evolutionary hypothesis, i.e., a hypothesis concerning a past event, it is difficult to test in current-living organisms. In principle, the hypothesis would predict that "social" organisms should be lateralized at the population level and "solitary" organisms at the individual level only. But this obviously refers to the conditions at the origin of a very complex evolutionary trajectory. For modern vertebrates arguing for completely solitary behavior is very difficult, at least in higher vertebrates (birds and mammals). Moreover, it is quite plausible that many current-living vertebrates that are considered today to be "solitary" actually derive from more social ancestors and therefore that they have retained population-level rather than individual-level asymmetries. Nonetheless a test of the hypothesis can be done using current-living species in which the distinction between solitary and social behavior can be defined quite clearly with respect to at least some aspects of behavior and in which it is likely that no major changes in their sociality have occurred in evolutionary terms.

A case in point is anti-predatory behavior of fish that shoal versus those that do not shoal. Shoaling in fish is a way of gaining protection against predators and it has been shown that this sort of grouping can arise from very simple "selfish" principles. Fishes can be easily categorized as "shoaling" or "not shoaling" species. We investigated whether shoaling in fish is associated with a population bias to turn in one direction (either left or right) when faced by a barrier of vertical bars through which a dummy predator could be seen.[94] The social tendency of the species was determined in terms of tendency to school: groups of fish were placed in a tank together and an index of their proximity to each other was determined. Six species were found to be gregarious (i.e., to school) and all six were the ones lateralized for turning bias at the population level; ten species were found to be nongregarious (i.e., not schooling) and six of these were not lateralized at the population level, but they were lateralized at the individual level. Thus, although the correlation is not perfect (and in biology it would be surprising if it were), the data fit our hypothesis quite well.

Further empirical evidence for the hypothesis of brain lateralization as an evolutionarily stable strategy arises from recent work on handedness in humans, that converges towards a crucial role of frequency-dependent selection. Humans exhibit hand preference for most manual activities with left-handers being a minority in all human populations.[1] The persistence of the polymorphism of handedness is a puzzle, because this trait is heritable (though its underlying genetic mechanisms are far from being clarified) and severe fitness costs are associated with left-handedness. Thus, some sort of benefit should exist to counteract these costs and to maintain the polymorphism. It has been found that left-handers have an advantage in sports involving dual confrontations, such as fencing, tennis and baseball, but not in noninteractive sports such as gymnastics.[95] Hence, this advantage does not arise from the well-known association between use of the left hand and direct control of it by the more visuo-spatial talented right-hemisphere. The advantage is frequency-dependent: left-handers have an advantage because they are relatively uncommon, as both left- and right-handers are less familiar with this category of competitor. Analyses of cricket have shown that the frequency of left-handers in this sport is best explained by a negative frequency-dependent selection mechanism.[96]

Faurie and Raymond[97] have recently put the hypothesis of a frequency-dependent advantage of left-handers in fights a step further, showing that the frequency of left-handers is strongly and positively correlated with the rate of homicides across traditional societies: ranging from 3% in the most pacifistic societies to 27% in the most violent and warlike. The interpretation of this finding would be that the advantage of being left-handed should be greater in a more violent context, which should result in a higher frequency of left-handers. In the absence of any selection pressure, the resulting equilibrium should be a 1:1 ratio of right-/left-handers, which has never been observed in any human population.

The last point also serves to clarify a theoretical issue. In my view, the problem of explaining the existence of left-handers (or of their equivalents in other species in other sensory-motor domains, such as e.g., biases in the use of one eye) is secondary with respect to the more general problem of explaining why lateralization, in several cases, does not conform to a 1:1 ratio of right- and left-forms. I believe that the appearance of a shift from the expected 1:1 ratio (i.e., the alignment of asymmetries at the level of population) is the result of an evolutionarily stable strategy and that the maintenance of a minority of individuals showing a different direction of lateralization (rather than a 100:0 distribution) is the indirect outcome of a frequency-dependent advantage exploited by this minority.

It is also important to stress that all this may occur only because brain asymmetries could manifest themselves in overt behavior as left-right biases. If asymmetries in the brain were without any effect in the left-right behavior of organisms, then no selection pressures for aligning the direction of asymmetries among different individuals would arise. Consider for instance the specialization of neural structures in the left side of the brain to control vocalization in birds (or for what matter even in humans or other species, review in Vallortigara and Rogers[89]). Assuming that these brain asymmetries do not produce any asymmetry in overt behavior, then no selection pressures for aligning them at the population level should have emerged. However, whether or not this is the case, the important point to stress is that when asymmetries that affect overt behavior are aligned under "social" pressures, then even those brain asymmetries that would be not constrained by social factors (because they do not have any obvious overt asymmetric manifestation in behavior) would nonetheless be forced to be located in the same hemisphere in all (or most) individuals of the species.

Alternative views to explain the riddle of directional lateralization have been proposed. For instance, Corballis[98] has argued for a heterozygotic advantage, maintaining balanced polymorphism. He refers to models which have been proposed to explain human handedness (see e.g., refs. 99,100), based on the presumed existence of a single gene locus, with one allele, D, specifying dextrality, and another, C, specifying chance. But there are problems with this hypothesis. One is that it is not clear at all why heterozygosity might be more adaptive than homozygosity in the case of brain lateralization. McManus[1] has presented a fascinating theory of **random cerebral variation** according to which beneficial combination of modules may occur more commonly in DC individuals, and in general an increased variability should be expected among left-handers. Yet, I do not see why such variability would necessarily produce an increase in biological fitness in individuals. Random variation may well produce an unbeneficial combination of cerebral modules.

The theory of the heterozygotic advantage also meets with difficulty with phenomena of oscillation in the relative frequency of different types of handedness over time periods, like those observed in scale-eating Cichlid fish[101] and with the evidence discussed above showing that left-handedness in humans is clearly associated with frequency-dependent selection. Thus, it seems to me (biased as I am) that the overall evidence tends to favor the ESS hypothesis over the heterozygotic advantage hypothesis.

One further fascinating test of the ESS hypothesis is that it predicts changes in the strength of occurrence of behavioral asymmetries depending on ecological factors, such as degree of predatory pressures. Evidence for this has been provided very recently by Brown et al[102] who found that populations of a single fish species collected from high predation areas show strongly lateralized behaviors whereas fish from low predation areas show no evidence of cerebral lateralization.[102] This study is very interesting because it is the first to contrast patterns of lateralization observed in wild populations; population differences in lateralized behaviors may be far more common than previously realized.

Several other issues deserve to be fully investigated in the near future. For instance, as I mentioned before, coordinated anti-predator responses provide an excellent example of positive frequency-dependent selection, since the behavior of each individual is reinforced by similar behaviors displayed by the majority of individuals in the group. An individual's fitness is dependent upon its ability to conform to the rest of the group. However, competition for food also increases with group size and therefore, in some circumstances, it may pay to behave differently from the rest of the group. Group foraging behavior, therefore, may provide us with an example of negative frequency-dependent selection. Obtaining the correct balance between foraging and anti-predator behavior is essential if individuals are to maximize fitness within the context of their contemporary environments. The balance between these two contrasting selective forces can predict and explain the variation in the pattern of lateralization we can observe at the population level in a wide range of species.

I envisage that, in the near future, lateralization research will move further and further from the laboratory of the neuropsychologist to the fieldwork of the behavioral neuroecologist.

Acknowledgements

Preparation of this manuscript was made possible by financial support provided by grants MIUR Cofin 2004, 2004070353_002 "Intellat" and MIPAF "Benolat" via Dip. Sci. Zootecniche Univ. of Sassari.

References

1. McManus IC. Right Hand, Left Hand. London: Widenfeld and Nicolson, 2002.
2. Vallortigara G. Comparative neuropsychology of the dual brain: A stroll through left and right animals' perceptual worlds. Brain Lang 2000; 73:189-219.
3. Vallortigara G, Rogers LJ, Bisazza A. Possible evolutionary origins of cognitive brain lateralization. Brain Res Brain Res Rev 1999; 30:164-175.
4. Rogers LJ, Andrew RJ. Comparative Vertebrate Lateralization. Cambridge: Cambridge University Press, 2002.
5. Tinbergen N. On aims and methods of ethology. Z Tierpsychol 1963; 20:410-433.
6. Weidner C, Reperant J, Miceli D et al. An anatomical study of ipsilateral retinal projections in the quail using autoradiographic, horseradish peroxidase, fluorescence and degeneration technique. Brain Res 1985; 340:99-108.
7. Rogers LJ, Deng C. Light experience and lateralization of the two visual pathways in the chick. Behav Brain Res 1999; 98:277-287.
8. Rogers LJ. Lateralization in vertebrates: Its early evolution, general pattern and development. In: Slater PJB, Rosenblatt J, Snowdon C, Roper T, eds. Advances in the Study of Behavior, Vol. 31. 2002:107-162.
9. Vallortigara G, Regolin L, Rigoni M et al. Delayed search for a concealed imprinted object in the domestic chick. Anim Cogn 1998; 1:17-24.
10. Regolin L, Rugani R, Pagni P et al. Delayed search for a social and a nonsocial goal object by the young domestic chick (Gallus gallus). Anim Behav 2005; in press.
11. Maguire EA, Frackowiack RSJ, Frith CD. Recalling routes around London: Activation of the right hippocampus in taxi drivers. J Neurosci 1997; 17:7103-7110.
12. Tommasi L, Gagliardo A, Andrew RJ et al. Separate processing mechanisms for encoding geometric and landmark information in the avian hippocampus. Eur J Neurosci 2003; 17:1695-1702.
13. Vallortigara G, Pagni P, Sovrano VA. Separate geometric and nongeometric modules for spatial reorientation: Evidence from a lopsided animal brain. J Cogn Neurosci 2004; 16:390-400.
14. Tommasi L, Vallortigara G. Encoding of geometric and landmark information in the left and right hemispheres of the avian brain. Behav Neurosci 2001; 115:602-613.
15. Vallortigara G. Visual cognition and representation in birds and primates. In: Rogers LJ, Kaplan G, eds. Vertebrate Comparative Cognition: Are Primates Superior to Nonprimates? New York: Kluwer Academic/Plenum Publishers, 2004:57-94.
16. Vallortigara G. The cognitive chicken: Visual and spatial cognition in a nonmammalian brain. In: Wasserman EA, Zentall TR, eds. Comparative Cognition: Experimental Explorations of Animal Intelligence. Oxford: Oxford University Press, 2005:in press.
17. Tommasi L, Vallortigara G, Zanforlin M. Young chickens learn to localize the centre of a spatial environment. J Comp Physiol [A] 1997; 180:567-572.
18. Güntürkün O. Avian cerebral asymmetries: The view from the inside. Cortex 2005; in press.
19. Folta K, Diekamp B, Güntürkün O. Asymmetrical modes of visual bottom-up and top-down integration in the thalamic nucleus rotundus of pigeons. J Neurosci 2004; 24:9475-9485.
20. Regolin L, Vallortigara G. Perception of partly occluded objects by young chicks. Percept Psychophys 1995; 57:971-976.
21. Forkman B. Hens use occlusion to judge depth in a two-dimensional picture. Perception 1998; 27:861-867.
22. Forkman B, Vallortigara G. Minimization of modal contours: An essential cross species strategy in disambiguating relative depth. Anim Cogn 1999; 4:181-185.
23. Regolin L, Marconato F, Vallortigara G. Hemispheric differences in the recognition of partly occluded objects by newly-hatched domestic chicks (Gallus gallus). Anim Cogn 2004; 7:162-170.
24. Corballis PM, Fendrich R, Shapley RM et al. Illusory contour perception and amodal boundary completion: Evidence of a dissociation following callosotomy. J Cogn Neurosci 1999; 11:459-466.
25. Vallortigara G, Andrew RJ. Lateralization of response by chicks to change in a model partner. Anim Behav 1991; 41:187-194.
26. Vallortigara G, Andrew RJ. Differential involvement of right and left hemisphere in individual recognition in the domestic chick. Behav Processes 1994; 33:41-58.
27. Dharmaretnam M, Andrew RJ. Age- and stimulus-specific use of right and left eyes by the domestic chick. Anim Behav 1994; 48:1395-1406.
28. Vallortigara G, Regolin L, Bortolomiol G et al. Lateral asymmetries due to preferences in eye use during visual discrimination learning in chicks. Behav Brain Res 1996; 74:135-143.
29. Gross Y, Franko R, Lewin I. Effects of voluntary eye movements on hemispheric activity and choice of cognitive mode. Neuropsychologia 1978; 17:653-657.

30. Deng C, Rogers LJ. Prehatching visual experience and lateralization in the visual Wulst of the chick. Behav Brain Res 2002; 134:375-385.
31. Rogers LJ. Development of lateralization. In: Andrew RJ, ed. Neural and Behavioral Plasticity: The Use of the Domestic Chick as a Model. Oxford: Oxford University Press, 1991:507-535.
32. Rogers LJ. Advantages and disadvantages of lateralization. In: Rogers LJ, Andrew RJ, eds. Comparative Vertebrate Lateralization. Cambridge: Cambridge University Press, 2002:126-153.
33. Rogers LJ. Light input and the reversal of functional lateralization in the chicken brain. Behav Brain Res 1990; 38:211-221.
34. Güntürkün O. The ontogeny of visual lateralization in pigeons. German Journal of Psychology 1993; 17:276-287.
35. Güntürkün O. Avian visual lateralization: A review. Neuroreport 1997; 8:3-11.
36. Güntürkün O. Hemispheric asymmetry in the visual system of birds. In: Hugdahl K, Davidson RJ, eds. Brain Asymmetry. 2nd ed. Cambridge: MIT Press, 2002:3-36.
37. Andrew RJ, Johnston ANB, Robins A et al. Light experience and the development of behavioral lateralization in chicks II. Choice of familiar versus unfamiliar model social partner. Behav Brain Res 2004; 155:67-76.
38. Vallortigara G, Cozzutti C, Tommasi L et al. How birds use their eyes: Opposite left-right specialisation for the lateral and frontal visual hemifield in the domestic chick. Curr Biol 2001; 11:29-33.
39. Deng C, Rogers LJ. Social recognition and approach in the chick: Lateralization and effect of visual experience. Anim Behav 2002; 63:697-706.
40. Rogers LJ, Andrew RJ, Burne THJ. Light exposure of the embryo and development of behavioral lateralization in chicks: I. Olfactory responses. Behav Brain Res 1998; 97:195-200.
41. Andrew RJ, Watkins JAS. Evidence of cerebral lateralization from senses other than vision. In: Andrew RJ, Rogers LJ, eds. Comparative Vertebrate Lateralization. Cambridge: Cambridge University Press, 2002:365-382.
42. Chiandetti C, Regolin L, Rogers LJ et al. Effects of light stimulation in embryo on the use of position-specific and object-specific cues in binocular and monocular chicks. Behav Brain Res 2005; in press.
43. Bobbo D, Galvani F, Mascetti GG et al. Light exposure of the chick embryo influences monocular sleep. Behav Brain Res 2002; 134:447-466.
44. Mascetti GG, Vallortigara G. Why do birds sleep with one eye open? Light exposure of chick embryo as a determinant of monocular sleep. Curr Biol 2001; 11:971-974.
45. Vallortigara G, Bisazza A. How ancient is brain lateralization? In: Rogers LJ, Andrew RJ, eds. Comparative Vertebrate Lateralization. 2002:9-69.
46. Bisazza A, Facchin L, Vallortigara G. Heritability of lateralization in fish: Concordance of right-left asymmetry between parents and offspring. Neuropsychologia 2000; 38:907-912.
47. Bisazza A, Sovrano VA, Vallortigara G. Consistency among different tasks of left-right asymmetries in lines of fish originally selected for opposite direction of lateralization in a detour task. Neuropsychologia 2001; 39:1077-1085.
48. Barth KA, Miklosi A, Watkins J et al. Fsi zebrafish show concordant reversal of laterality of viscera, neuroanatomy and a subset of behavioral responses. Curr Biol 2005; in press.
49. Gamse JT, Thisse C, Thisse B et al. The parapineal mediates left-right asymmetry in the zebrafish diencephalons. Development 2003; 130:1059-1068.
50. Concha ML, Burdine RD, Russell C et al. A nodal signaling pathway regulates the laterality of neuroanatomical asymmetries in the zebrafish forebrain. Neuron 2000; 28:399-409.
51. Halpern ME, Liang JO, Gamse JT. Leaning to the left: Laterality in the zebrafish forebrain. Trends Neurosci 2003; 26:308-313.
52. Hauser MD. Right hemisphere dominance for the production for the production of facial expression in monkeys. Science 1993; 261:475-477.
53. Hook-Costigan MA, Rogers LJ. Lateralization of hand, mouth and eye use in the common marmoset (Callithrix jacchus). Folia Primatologica 1995; 64:180-191.
54. Bradshaw JL, Nettleton NC. Human Cerebral Asymmetry. Englewood Cliff: Prentice Hall, 1983.
55. De Renzi E, Perani D, Carlesimo GA et al. Prosopagnosia can be associated with damage confined to the right hemisphere: An MRI and PET study and a review of the literature. Neuropsychologia 1994; 179:893-902.
56. Perrett DI, Rolls ET, Caan W. Visual neurones responsive to faces in the monkey temporal cortex. Exp Brain Res 1982; 47:329-342.
57. Kendrick KM, Baldwin BA. Cells in the temporal cortex of sheep can respond preferentially to the sight of faces. Science 1987; 236:448-450.

58. Kendrick KM. Brain asymmetries for face recognition and emotion control in sheep. Cortex 2005; in press.
59. Bisazza A, Cantalupo C, Robins A et al. Right-pawedness in toads. Nature 1996; 379:408.
60. Malaschichev YB, Wassersug RJ. Left and right in the amphibian world: Which way to develop and where to turn? BioEssays 2004; 26:512-522.
61. Rogers LJ. Lateralised brain function in anurans: Comparison to lateralization in other vertebrates. Laterality 2002; 7:219-239.
62. Vallortigara G, Rogers LJ, Bisazza A et al. Complementary right and left hemifield use for predatory and agonistic behavior in toads. NeuroReport 1998; 9:3341-3344.
63. Diekamp B, Regolin L, Güntürkün O et al. A left-sided visuospatial bias in birds. Curr Biol 2005; 15:R372-R373.
64. Robins A, Lippolis G, Bisazza A et al. Lateralised agonistic responses and hind-limb use in toads. Anim Behav 1998; 56:875-881.
65. Rogers LJ. Light experience and asymmetry of brain function in chickens. Nature 1982; 297:223-225.
66. Deckel AW. Lateralization of aggressive responses in Anolis. J Exp Zool 1995; 272:194-200.
67. Deckel AW. Effects of alcohol consumption on lateralized aggression in Anolis carolinensis. Brain Res 1997; 756:96-105.
68. Casperd LM, Dunbar RIM. Asymmetries in the visual processing of emotional cues during agonistic interactions by gelada baboons. Behav Processes 1996; 37:57-65.
69. Lippolis G, Bisazza A, Rogers LJ et al. Lateralization of predator avoidance responses in three species of toads. Laterality 2002; 7:163-183.
70. Ventolini N, Ferrero E, Sponza S et al. Laterality in the wild: Preferential hemifield use during predatory and sexual behavior in the Black winged stilt (Himantopus himantopus). Anim Behav 2005; 69:1077-1084.
71. Milinski M. TIT FOR TAT in sticklebacks and the evolution of cooperation. Nature 1987; 325:433-435.
72. Bisazza A, De Santi A, Vallortigara G. Laterality and cooperation: Mosquitofish move closer to a predator when the companion is on their left side. Anim Behav 1999; 57:1145-1149.
73. Sovrano VA. Visual lateralization in response to familiar and unfamiliar stimuli in fish. Behav Brain Res 2004; 152:385-391.
74. Sovrano VA, Bisazza A, Vallortigara G. Lateralization of response to social stimuli in fishes: A comparison between different methods and species. Physiol Behav 2001; 74:237-244.
75. Sovrano V, Rainoldi C, Bisazza A et al. Roots of brain specializations: Preferential left-eye use during mirror-image inspection in six species of teleost fish. Behav Brain Res 1999; 106:175-180.
76. De Santi A, Bisazza A, Cappelletti M et al. Prior exposure to a predator influences lateralization of cooperative predator inspection in the guppy, Poecilia reticulata. Ital J Zool 2000; 67:175-178.
77. De Santi A, Sovrano VA, Bisazza A et al. Mosquitofish display differential left- and right-eye use during mirror-image scrutiny and predator-inspection responses. Anim Behav 2001; 61:305-310.
78. Levy J. The mammalian brain and the adaptive advantage of cerebral asymmetry. Ann N Y Acad Sci 1977; 299:264-272.
79. Andrew RJ. The nature of behavioral lateralization in the chick. In: Andrew RJ, ed. Neural and behavioral Plasticity. The Use of the Chick as a Model. Oxford: Oxford University Press, 1991:536-554.
80. Andrew RJ. Behavioral development and lateralization. In: Rogers LJ, Andrew RJ, eds. Comparative Vertebrate Lateralization. Cambridge: Cambridge University Press, 2002:157-205.
81. Andrew RJ, Mench J, Rainey C. Right-left asymmetry of response to visual stimuli in the domestic chick. In: Ingle DJ, Goodale MA, Mansfield RJ, eds. Analysis of Visual Behavior. Cambridge: MIT Press, 1982:225-236.
82. Cantalupo C, Bisazza A, Vallortigara. Lateralization of predator-evasion response in a teleost fish (Girardinus falcatus). Neuropsychologia 1995; 33:1637-1646.
83. Rogers LJ, Zucca P, Vallortigara G. Advantages of having a lateralized brain. Proc Biol Sci 2004; 271:420-422.
84. Rogers LJ. Evolution of hemispheric specialisation: Advantages and disadvantages. Brain Lang 2000; 73:236-253.
85. Denenberg VH. Hemispheric laterality in animals and the effects of early experience. Behav Brain Sci 1981; 4:1-49.
86. Collins RL. On the inheritance of direction and degree of asymmetry. In: Glick SD, ed. Cerebral Lateralization in Nonhuman Species. New York: Academic Press, 1985:41-71.
87. Heuts BA. Lateralization of trunk muscle volume, and lateralization of swimming turns of fish responding to external stimuli. Behav Processes 1999; 47:113-124.

88. McGrew WC, Marchant LF. Laterality of hand use pays off in foraging success for wild chimpanzees. Primates 1999; 40:509-513.
89. Vallortigara G, Rogers LJ. Survival with an asymmetrical brain: Advantages and disadvantages of cerebral lateralization. (Target article) Behav Brain Sci 2005; 28(4):575-633.
90. Rogers LJ. Laterality in animals. Int J Comp Psychol 1989; 3:5-25.
91. Maynard-Smith J. Evolution and the theory of games. Cambridge: Cambridge University Press, 1982.
92. Ghirlanda S, Vallortigara G. The evolution of brain lateralization: A game theoretical analysis of population structure. Proc Biol Sci 2004; 271:853–57.
93. Foster WA, Treherne JE. Evidence for the dilution effect in the selfish herd from fish predation of a marine insect. Nature 1981; 293:508-510.
94. Bisazza A, Cantalupo C, Capocchiano M et al. Population lateralization and social behavior: A study with sixteen species of fish. Laterality 2000; 5:269-284.
95. Raymond M, Pontier D, Dufour A et al. Frequency-dependent maintenance of left handedness in humans. Proc Biol Sci 1996; 263:1627-1633.
96. Brooks R, Bussière LF, Jennions MD et al. Sinister strategies succeed at the cricket World Cup. Proc Biol Sci 2004; 271:S64-S66.
97. Faurie C, Raymond M. Handedness, homicide and negative frequency-dependent selection. Proc Biol Sci 2005; 272:25-28.
98. Corballis MC. The genetics and evolution of handedness. Psychol Rev 1997; 104:714-727.
99. Annett M. The right shift theory of a genetic balanced polymorphism for cerebral dominance and cognitive processing. Curr Psychol Cognit 1995; 14:427-480.
100. McManus IC, Bryden MP. The genetics of handedness and cerebral lateralization. In: Rapin I, Segalowitz SJ, eds. Handbook of Neuropsychology, Vol. 6. Amsterdam: Elsevier, 1992:115-144.
101. Hori M. Frequency-dependent natural selection in the handedness of scale-eating Cichlid fish. Science 1993; 260:216-219.
102. Brown C, Gardner C, Braithwaite VA. Population variation in lateralized eye use in the poeciliid Brachyraphis episcope. Proc Biol Sci 2004; 271(Suppl 6):S455-457.
103. Regolin L, Garzotto B, Rugani R et al. Working memory in the chick: Parallel and lateralized mechanisms for encoding of object- and position-specific information. Behav Brain Res 2004; 157:1-9.

CHAPTER 10

Cognitive and Social Advantages of a Lateralized Brain

Lesley Rogers*

Abstract

Of the many examples of lateralization in vertebrates some are expressed at the individual level only (i.e., not aligned in the population) and others at both the individual and population level. This chapter addresses the advantages and disadvantages of both manifestations of lateralization. First, it discusses results of experiments conducted with chicks and marmosets showing that having a lateralized brain enhances an animal's ability to perform more than one task simultaneously. By allocating the processing required for one task (searching for food) to the left hemisphere and that required for the other task (detecting a predator) to the right hemisphere, animals increase their capacity to attend to both tasks at the same time. Since this advantage of having a lateralized brain applies only to the individual and does not require lateralization at the population level, another explanation is needed for the latter. Indeed, population level lateralization would seem to have the disadvantage of, for example, predators exploiting their prey's bias to respond to their presence more readily on the left side. Hence, this apparent disadvantage might have to be counteracted by other distinct advantages of population lateralization. Here the hypothesis that advantages occur in social interactions between lateralized individuals is considered. Some concluding remarks are made about lateralization in primates, and its potential association with social behavior, and the development of lateralization in the chick as a model demonstrating the multiple interactive influences on lateralization.

Introduction

Lateralization of brain function is wide spread amongst vertebrates.[1,2] Some expressions of lateralization occur at the individual level without the presence of a directional bias in the population, and others at both the individual and the population level. When considering the potential advantage of having a lateralized brain, we should treat these two forms of lateralization separately since different selective pressures are likely to impinge on each type. Whereas it may be advantageous for an individual to be lateralized, the alignment of the direction of lateralization in most members of a population would seem to have clear disadvantages. For example, a population bias to be more responsive to predators advancing on the left than on the right could be exploited by a predator. Nevertheless such a directional bias has been found in several species (toads;[3] dunnarts, a marsupial species;[4] chicks[5]). Hence, it seems that advantages must ensue from not only the individual having a lateralized brain but also from the majority of individuals being lateralized in the same direction.

*Lesley Rogers—Centre for Neuroscience and Animal Behavior, Building W28, University of New England, Armidale, NSW 2351, Australia. Email: lrogers@une.edu.au

Behavioral and Morphological Asymmetries in Vertebrates, edited by Yegor B. Malashichev and A. Wallace Deckel. ©2006 Landes Bioscience.

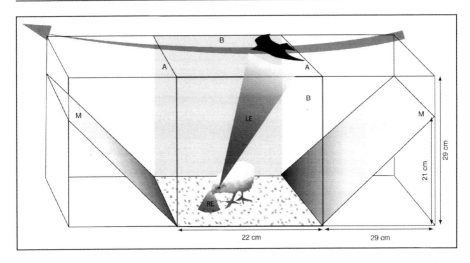

Figure 1. Testing apparatus for the dual task, requiring the chick to peck for grains of mash scattered on a background of small pebbles adhered to the floor while at the same time monitoring for the silhouette of a model predator moved overhead. Note the two mirrors (M), which allowed us to see the chick from the sides in a video filmed from overhead. The sides of the cage marked B are opaque, and those marked A transparent. The targets of the chick's pecks (grain versus pebbles) were scored by direct observation through a window at floor level. The responses to the predator were scored by replaying the videotape. LE: left eye; RE: right eye.

What Is the Advantage of the Individual Having a Lateralized Brain?

For some time, it has been assumed that brain lateralization is an advantage because it increases neural capacity.[6,7] It is said to achieve this by engaging only one hemisphere in the performance of a particular function, thus leaving the other hemisphere free to perform another function. Until recently, there has been no convincing evidence in support of this hypothesis, but now we have shown that a chick with a lateralized brain can perform two tasks that engage the opposite hemispheres simultaneously more efficiently than a chick not lateralized for performing these tasks.[8]

The tasks chosen for testing the chicks were: (1) foraging for grain against a background of small pebbles adhered to the floor (the pebble floor test)—this task uses the left hemisphere[9] and relies on inputs from the right eye,[10,11] and (2) detecting the silhouette of a predator overhead—this task is performed primarily using the left eye and right hemisphere.[5,12] [Note that the visual input to each eye is processed primarily, but not exclusively, by the contralateral hemisphere (more on this in Chapter 5 by Rogers and Kaplan in this volume)].

The testing apparatus is shown in Figure 1. The model predator was a shape cut from black card and it resembled the silhouette of a hawk (8 cm wing span). It was moved over the cage (across its longest length) at a speed of 12 cm per second every 18 seconds. This was achieved by attaching the predator to one end of a fine metal rod and the other end to a motor. The apparatus was designed to test the chick's visual responses to a model predator, and not auditory responses. The presentation of the predator was not accompanied by any auditory cues and ambient sounds were not lateralized: the motor providing movement to the stimulus was placed at the midpoint of the cage and was running constantly (i.e., it provided a nonlateralized, white noise). The absence of auditory cues affecting the results was also confirmed in a later repeat of the results reported here using a silent video image of a hawk presented overhead instead of the mechanically presented silhouette of the hawk.[13]

Table 1. **Performance on the pebble-floor task by chicks tested binocularly or monocularly**

Eye Condition	Predator Presented	Light Exposed	Dark-Incubated
Binocular	No	+	+
Binocular	Yes	+	--
Left eye open	Yes	-	-
Right eye open	Yes	+	-

+: good performance—able to avoid pecking at pebbles; -: poor performance—not able to avoid pecking pebbles; --: performance deteriorates as the task progresses

The latter experiment was conducted in an insulated room that was largely sound proof. Since this experiment gave the same pattern of results as outlined below, we can be sure that the chick is responding to visual cues only.

In the tests conducted by Rogers et al[8] the chicks were deprived of food for several hours before being tested. Once a chick had pecked at grain of pebbles five times, the predator was presented for the first time. The presentation was repeated until the chick had completed 65 pecks. Performance on the pebble-floor task and responsiveness to the predator were scored.

The performance of chicks hatched from eggs incubated in the dark was compared to that of chicks hatched from eggs exposed to light during the final stages of incubation. The light-exposed chicks have lateralization for performance of the two tasks (left hemisphere for the pebble-floor task and right-hemisphere for detection of the predator), whereas the dark-incubated chicks have no lateralization of these functions.[5,10,11]

Exposure of the eggs to light before hatching establishes the lateralization as a consequence of the embryo's orientation within the eggshell. The late-stage embryo occludes its left eye with its body, whereas its right eye can be stimulated by light entering the egg through the shell and membranes. This postural monocularity occurs at a stage of development when the visual connections to the pallium (forebrain) are becoming functional[14] and it stimulates better development of the visual projections from the right eye compared to those from the left eye.[15,16] As a consequence, asymmetry develops in the thalamofugal pathway but, incidentally, not in the tectofugal pathway.[17]

Hence, light exposure of the eggs just prior to hatching leads to lateralization of the thalamofugal visual pathway and of the two patterns of behavior tested in the paradigm described. Incubation of the eggs in the dark prevents the development of these lateralities. At least in the case of the pebble-floor task we know that the thalamofugal visual pathway has a critical role, as shown by injection of glutamate into various regions of the pallium.[9,18]

The light-exposed chicks learnt to avoid pecking pebbles both with and without presentation of the predator, and they retained a memory of the task on the next day.[8] The dark-incubated chicks were unable to learn to avoid pebbles when the predator was present and, in fact, they pecked at pebbles more often as the task progressed. In other words, their ability to find grain against the distracting background of pebbles deteriorated as the task progressed. They also showed poor memory of the task on the next day. However, the dark-incubated chicks had no difficulty in pecking at the grain and avoiding the pebbles when no predator was presented (results summarized in Table 1).

In addition to their poor performance of the pebble-floor task, the dark-incubated chicks were more likely to fail to respond to the predator as it passed overhead. They often continued to peck during its presentation, whereas the light-exposed chicks interrupted pecking and attended to the stimulus overhead, at least briefly. They viewed the predator with the left eye, often circling so that they could continue to do so, or they simply opted immediately for the avoidance response of crouching, often preceded by running steps.

Most chicks used their left eye for sustained viewing of the predator stimulus as it moved across the cage but, if a light-exposed, lateralized chick happened to catch sight of the predator with its right eye, it would shift to use its left eye immediately and maintain fixation with this eye.[13] Dark-incubated chicks showed no preferred eye for viewing the predator, which is confirmation of their lack of lateralization for this task.

This result shows that lateralized chicks can perform the two tasks simultaneously, likely because one demands use of the left hemisphere and the other use of the right hemisphere. Without the discrete allocation of these functions to opposite hemispheres, performance is impaired and becomes increasingly so as the dual demands continue. Nonlateralization, therefore, appears to cause a state of confusion or task interference.

Since the nonlateralized chicks perform the pebble-grain task as well as the lateralized ones provided no predator is presented, they are not merely impaired in some general sense but only when dual, and presumably also multiple, demands are made. One could argue that the dark-incubated, nonlateralized chicks were simply more disturbed by presentation of the predator than were the light-exposed, lateralized ones. The answer to this is not straightforward since degree of responsiveness to the predator cannot be clearly separated from the increasing confusion that nonlateralized chicks experience in the dual task. The fact that nonlateralized chicks were less likely to respond to, or detect, the predator than were the lateralized chicks might indicate that they were less fearful of the predator. However, although they did not show higher levels of reactivity measured in terms of crouching, running away or startle response, they did distress peep more than lateralized chicks and they took longer to resume pecking once they had been interrupted.[13] Although the latter could be interpreted as an indication of higher levels of distress caused by the presence of the predator, it could also result from the inability of the nonlateralized chicks to find grain amongst the pebbles.

Dharmaretnam and Rogers[13] investigated the role of lateralization in performance of the two tasks simultaneously by testing the chicks monocularly and revealed that the superior ability of light-exposed, lateralized chicks to find grain amongst the pebbles was due to a specific effect of light exposure on the development of the visual pathways receiving input from the right eye (Table 1). Once the embryo's right eye has been exposed to light, it seems, that eye system is put in charge of foraging, whereas response to predators, and/or novel stimuli, is allocated to the visual systems fed by the left eye.

To examine the generality of this proposed function of brain lateralization, we have recently tested marmosets, *Callithrix jacchus*, using a paradigm similar to the one used for chicks (Piddington and Rogers, in preparation). Lateralization strength in the marmosets was determined as strength of hand preference for picking up pieces of food and holding them to the mouth during eating (details in ref. 19; Fig. 2). The marmosets were trained to search for favored food (mealworms) placed in small blue pots (one worm per pot and eight pots) attached to branches in a room (Fig. 3). A green pot was attached next to each blue pot but not baited with food. Once trained, the marmosets were tested. Each marmoset was allowed to enter the room and, once it had obtained one mealworm, a predator was introduced, either a hawk moved across the room just below the ceiling or a snake moved across the floor. The hawk was a taxidermic specimen and the snake was a rubber model. They were moved mechanically using nylon line and a pulley system. We found a strong positive correlation between the strength of hand preference (regardless of whether the left of right hand was preferred) and the latency to detect the predator; the weaker the lateralization, the longer the latency (Fig. 4). When the same marmosets were tested on the single task of detecting the predator without the combined foraging task, no such association between lateralization and latency was found. Hence, as in chicks, marmosets that are less strongly lateralized, or not lateralized at all, are less efficient in detecting the predator. There was one difference between the performance of the marmosets and the

Figure 2. A marmoset holding food to the mouth. Hand preferences were determined by scoring 100 such events. Usually the marmoset was on the floor, not clinging to the wire cage wall.

chicks: the marmosets' success in finding mealworms was not associated with strength of lateralization, whereas in chicks finding food was associated with strength of lateralization. Likely, this difference resulted because the marmosets were well trained to perform the food searching aspect of the dual task prior to testing, whereas the chicks were learning during testing.

It is important to state here that the superior ability of lateralized animals on the dual task would be relevant to behavior in the natural environment. Rarely would the attention of wild animals be focused on performance of just one task since this would leave them open to predation and other risks. Hence, lateralization would have a clear selective advantage. Moreover, some environments might well place more demands on individuals than other environments and selection for lateralization would be stronger. One might imagine that this would be the case in habitats with high levels of predation. In fact, Brown et al[20] have provided evidence of this in fish: members of a species of poeciliid fish were found to exhibit lateralized eye use if they were in a habitat with high levels of predation and they exhibited no lateralization if they were in a habitat with low levels of predation.

Figure 3. The dual task used to test marmosets. Blue bowls (light gray) baited with one mealworm each. Green bowls (dark gray), not baited. The hawk was presented overhead by pulling it on a nylon line from a box in which it had been concealed and then moving it across the ceiling of the room. The snake was presented in a separate trial and it was moved across the floor also on a nylon line. See text for details. Piddington and Rogers, in preparation.

What Is the Advantage of the Population Being Lateralized?

Although the above experiments have shown the advantage of carrying out, in separate hemispheres, the functions of predator detection and finding food, the same advantage could be achieved irrespective of the direction of the lateralization. Half of the individuals could be lateralized in one direction and half in the other without effect on this advantage of lateralization. However, most of the types of lateralization that I have discussed so far are lateralized in the same direction and this could be disadvantageous. For example, a predator might exploit its prey's bias of detecting it less readily if it approaches on the prey's left side. Although this strategy of predators is given here as a possibility only, there is some evidence that predatory birds are lateralized, in foot preferences at least.[21]

It seems, therefore, that some clear advantages must counteract the potential disadvantages of directional bias at the population level. With this in mind, Giorgio Vallortigara and I have postulated that this advantage lies in predictability of social behavior amongst conspecifics.[22] As an example, we know that the right hemisphere of the chick activates attack responses and that this is manifested as a left-eye bias for directing agonistic pecks at unfamiliar conspecifics.[23] One chick might, therefore, avoid being attacked if it approaches another chick on the latter's right side. In fact, groups of light-exposed chicks (8 per group) form more rigid, or stable, social hierarchies than do groups of chicks hatched from eggs incubated in the dark.[24]

Hence, whether or not a type of lateralization has a role in social interactions may determine its directionality in the population. Within a species some lateralities are present at the individual level only, as is the case for hand preferences to pick up food in common marmosets,[19] and others at both the individual and population levels, as for eye preferences to view stimuli[25] and for facial expressions[26] in marmosets. Consistent with our hypothesis, facial

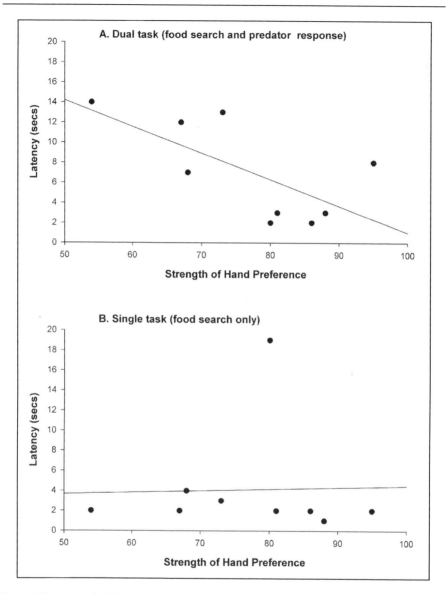

Figure 4. A) An example of the negative correlation between the marmoset's strength of hand preference and latency to detect the predator, in this case the hawk, introduced while the marmoset was searching for its favorite food (i.e., on the dual task). Note that animals with a weaker hand preference take longer to detect the introduced predator. B) The same animals tested with presentation of the hawk but this time without the simultaneous task of searching for food (i.e., on a single task). In this case there is no relationship between hand preference and latency to detect the predator.

expressions have a clear social function in communication, whereas hand preferences for holding food may not. If our hypothesis is correct, eye preference should have some kind of function in social behavior but this is not known for the marmoset. It is, however, known that orang-utans show a side preference, and thus an eye preference, to look sideways at other individuals, this being their main mode of social gazing.[27]

Concluding Remarks

The examples discussed above demonstrate the importance of assessing lateralization of a range of functions within a species and the need to investigate further the hypothesized association between directional bias and social behavior. The presence of lateralization in primates has been, and still is, highly controversial and it has lead to division between those who believe that lateralization is a uniquely human trait[28] and those of us who are studying lateralization in nonhuman vertebrates (summarized in ref. 29). For example, some researchers claim that apes lack a population bias in hand preferences[30] and others that a population bias is present.[31] One path to solving these apparent conflicts might be to select behavior that is clearly a part of social behavior and then look to see whether it is lateralized. In other words, the focus on hand preferences may well have been very misleading, especially in the case of feeding and even obtaining food using tools, since these behavior patterns may not involve social interaction in any specific way. Other patterns of behavior (e.g., direction of social gazing, touching of conspecifics, carrying of infants) are, I believe, more likely candidates for the expression of lateralization. Of course, other factors such as task complexity may also influence the expression of lateralization, as others have noted.[32]

Finally, it is worth returning to the chick model to consider the processes involved in the development of lateralization. As I have discussed elsewhere[33,34] the interactive roles of gene expression, circulating levels of steroid hormones and light stimulation of the chick embryo on the later manifestation of lateralization, here I will mention only the more general point that the development of lateralization seems to conform to a channelling process with multiple influences constraining development to the expression of both perceptual and motor lateralization. I have discussed some of the perceptual asymmetries in the chick visual system, others occur in olfactory[35] and auditory[36] perception. Chicks also display motor asymmetries: they show a population-level, right-foot preference for scraping the ground while feeding[37,38] and for the first foot to step off from a standing position.[39] They also display left-side turning preferences.[39,40,41] However, it remains unclear whether these are solely motor asymmetries or motor expressions of perceptual lateralization: at least for scratching while feeding, the preferred foot seems to be secondary to preferred use of the right eye in guiding pecking at food targets, as shown by monocular testing.[42] Nevertheless, it may well be irrelevant, or impossible, to decide whether a particular expression of lateralization is motor or perceptual since the chick appears to be characterized by an integrated symphony of lateralizations. Moreover, a number of embryonic events make up an integrated overture to the post-hatching expression of lateralization.

Not only does the orientation of the embryo in the egg, most likely determined by gene expression, lead to the asymmetrical stimulation of the eyes by light (discussed above) but also it must foreground the anticlockwise turning of the embryo as it breaks open the eggshell (Fig. 5), driven by the right side of the chick's body as it uses its limbs, head and neck to rotate within the egg.[43,44] The latter in turn influences the chick's expression of lateralization of motor behavior since disruption of the hatching process prevents the development of motor lateralization.[39,45] Added to this, visual experience prior to hatching influences not only perceptual lateralization (discussed above) but also motor turning; chicks given extra experience of patterned visual input by removal of the shell over the air sac just prior to hatching have altered turning behavior after hatching.[40] Hence, the late stage chick embryo relies on body posture, visual experience and its own motor activity during hatching to lead on to the expression of lateralization after hatching. The disruption of any one of these critical contributions to the canalizing of development precludes the expression of a broad range of motor and visual lateralizations. Now we need to know more about the effects of such disruptions on individual versus population level lateralization, and about the effects, if any, of these procedures on lateralization in the other perceptual modalities.

Furthermore, we might be encouraged to investigate the influence of events that take place earlier in the incubation period on these critical steps in late incubation. A sequence of

Figure 5. The asymmetrical position adopted by the chick embryo during the final days of incubation. One arrow shows the direction of turning as the embryo breaks open the shell during hatching. The other arrow indicates that light stimulates the right eye of the embryo, the left eye being occluded by the embryo's body.

positions adopted by the embryo leads to asymmetrical turning of the head during the final stages of incubation,[46] and incorrect positioning of the embryo can result from insufficient turning of eggs, preincubation egg storage and orientation of the egg,[47] the first two of these being of likely relevance to the hen's incubation of eggs in natural conditions. The sequential development of the sensory systems (summarized by ref. 14) may well be integrated with these motor positions and be part of the unfolding lateralized development of the nervous system. Species comparison of the hatching process within the framework of lateralization also awaits investigation and should prove worthwhile since there are some differences between species in the embryo's orientation and the hatching process.[48,49] In fact, Casey and Sleigh[50] have already demonstrated that Japanese quail do not show a population level turning preference after hatching and during hatching they do not rotate a full turn, as do domestic chicks, bantam chicks and bobwhite quail, all of the latter showing turning biases. Those species that hatch from the egg after making a longitudinal split in the shell (see Fig. 3 of ref. 51), and presumably without rotating, would be interesting to examine for motor and perceptual lateralization.

References

1. Rogers LJ. Lateralization in vertebrates: Its early evolution, general pattern and development. In: Slater PJB, Rosenblatt J, Snowdon C et al, eds. Advances in the Study of Behavior 2002; 31:107-162.
2. Rogers LJ, Andrew RJ. Comparative Vertebrate Lateralization. Cambridge: Cambridge University Press, 2002.
3. Lippolis G, Bisazza A, Rogers LJ et al. Lateralization of predator avoidance responses in three species of toads. Laterality 2002; 7:163-183.
4. Lippolis G, Westman W, McAllan BM et al. Lateralization of escape responses in the striped-faced dunnart, Sminthopsis macroura (Dasyuridae: Marsupalia). Laterality 2005; in press.

5. Rogers LJ. Evolution of hemispheric specialisation: Advantages and disadvantages. Brain Lang 2000; 73:236-253.

6. Dunaif-Hattis J. Doubling the Brain. New York: Peter Lang, 1984.

7. Levy J. The mammalian brain and the adaptive advantage of cerebral asymmetry. Ann New York Acad Sci 1977; 299:264-272.

8. Rogers LJ, Zucca P, Vallortigara G. Advantage of having a lateralized brain. Proc Royal Soc Lond B 2004; 271:S420-S422.

9. Deng C, Rogers LJ. Differential contributions of the two visual pathways to functional lateralization in chicks. Behav Brain Res 1997; 87:173-182.

10. Rogers LJ. Light input and the reversal of functional lateralization in the chicken brain. Behav Brain Res 1990; 38:211-221.

11. Rogers LJ. Early experiential effects on laterality: Research on chicks has relevance to other species. Laterality1997; 2:199-219.

12. Evans CS, Evans L, Marler P. On the meaning of alarm calls: Functional references in an avian vocal system. Anim Behav 1993; 46:23-28.

13. Dharmaretnam M, Rogers LJ. Hemispheric specialization and dual processing in strongly versus weakly lateralized chicks. Behav Brain Res 2005; 162:32-70.

14. Rogers LJ. The Development of Brain and Behavior in the Chicken. Wallingford: CAB International, 1995.

15. Koshiba M, Nakamura S, Deng C et al. Light-dependent development of asymmetry in the ipsilateral and contralateral thalamofugal visual projections of the chick. Neurosci Letts 2003; 336:81-84.

16. Rogers LJ, Sink HS. Transient asymmetry in the projections of the rostral thalamus to the visual hyperstriatum of the chicken, and reversal of its direction by light exposure. Exp Brain Res 1988; 70:378-384.

17. Rogers LJ, Deng C. Light experience and lateralization of the two visual pathways in the chick. Behav Brain Res 1999; 98:277-287.

18. Deng C, Rogers LJ. Prehatching visual experience and lateralization of the visual Wulst. Behav Brain Res 2002; 134:375-385.

19. Hook MA, Rogers LJ. Development of hand preferences in marmosets (Callithrix jacchus) and effects of ageing. J Comp Psychol 2000; 114:263-271.

20. Brown C, Gardner C, Braithwaite V. Population variation in lateralised eye use in the poeciliid Brachyraphis episcopi. Proc Royal Soc Lond B 2004; 271:S455-S457.

21. Csermely D. Lateralization in birds of prey: Adaptive and phylogenetic considerations. Behav Processes 2004; 67:511-520.

22. Vallortigara G, Rogers LJ. Survival with an asymmetrical brain: Advantages and disadvantages of cerebral lateralization. Behav Brain Sci 2005; 28(4):575-633.

23. Vallortigara G, Cozzutti C, Tommasi L et al. How birds use their eyes: Opposite left-right specialisation for the lateral and frontal visual hemifield in the domestic chick. Current Biol 2001; 11:29-33.

24. Rogers LJ, Workman L. Light exposure during incubation affects competitive behavior in domestic chicks. App Anim Behav Sci 1989; 23:187-198.

25. Hook-Costigan MA, Rogers LJ. Lateralized use of the mouth in production of vocalizations by marmosets. Neuropsychologia 1998; 36:1265-1273.

26. Hook-Costigan MA, Rogers LJ. Eye preferences in common marmosets (Callithrix jacchus): Influence of age, stimulus and hand preference. Laterality1998; 3:109-130.

27. Kaplan G, Rogers LJ. Patterns of gazing in orang-utans (Pongo pygmaeus). Int J Primatol 2002; 23:501-526.

28. Crow TJ. The speciation of modern Homo sapiens. Proc British Academy. Oxford: Oxford University Press, 2002:106:179-216.

29. McManus C. Review of Comparative Vertebrate Lateralization. Ann Human Biol 2004; 31:4-5.

30. Marchant LF, McGrew WC. Laterality of limb function in wild chimpanzees of Gombe National Park: Comprehensive study of spontaneous activities. J Human Evol 1996; 30:427-443.

31. Hopkins WD, Hook M, Braccini S et al. Population-level right handedness for a coordinated bimanual task in chimpanzees: Replication and extension in a second colony of apes. Int J Primatol 2003; 24:677-689.

32. Hopkins WD, Cantalapo C. Individual and setting differences in the hand preferences of chimpanzees (Pan troglodytes): A critical analysis and some alternative explanations. Laterality 2005; 10:65-80.

33. Deng C, Rogers LJ. Factors affecting the development of lateralization in chicks. In: Rogers LJ, Andrew RJ, eds. Comparative Vertebrate Lateralization. Cambridge: Cambridge University Press, 2002:206-246.

34. Rogers LJ. Factors influencing development of lateralization. Cortex 2005; in press.
35. Rogers LJ, Andrew RJ, Burne THJ. Light exposure of the embryo and development of behavioral lateralisation in chicks: I. Olfactory responses. Behav Brain Res 1998; 97:195-200.
36. Howard KJ, Rogers LJ, Boura ALA. Functional lateralisation of the chicken forebrain revealed by use of intracranial glutamate. Brain Res 1980; 188:369-382.
37. Dharmaretnam M, Vijitha V, Priyadharshini K et al. Ground scratching and preferred leg use in domestic chicks: Changes in motor control in the first two weeks post-hatching. Laterality 2002; 7:371-380.
38. Rogers LJ, Workman L. Footedness in birds. Anim Behav 1993; 45:409-411.
39. Casey MB, Martino C. Asymmetrical hatching behaviors influence the development of postnatal laterality in domestic chicks (Gallus gallus). Dev Psychobiol 2000; 34:1-12.
40. Casey MB, Karpinski S. The development of postnatal turning bias is influenced by prenatal visual experience in domestic chicks (Gallus gallus). Psych Record 1999; 49:67-74.
41. Casey MB, Lickliter R. Prenatal visual experience influences the development of turning bias in bobwhite quail chicks (Colinus virginianus). Dev Psychobiol 1998; 32:327-338.
42. Tommasi L, Vallortigara G. Footedness in binocular and monocular chicks. Laterality 1999; 4:89-95.
43. Bekoff A, Kauer JA. Neural control of hatching: Fate of the pattern generator for the leg movements of hatching in post-hatching chicks. J Neurosci 1984; 4:2659-2666.
44. Kuo ZY. The dynamics of behavior development: An epigenetic view. New York: Random House, 1967.
45. Casey MB. Developmental systems, evolutionarily stable strategies, and population laterality. Behav Brain Sciences 2005; in press.
46. Freeman BM, Vince MA. Development of the Avian Embryo. New York: John Wiley and Sons, 1994.
47. Wilson HR, Neuman SL, Eldred AR et al. Embryonic malpositions in broiler chickens and bobwhite quail. J App Poultry Res 2003; 12:14-23.
48. Oppenheim RW. Prehatching and hatching behavior in birds: A comparative study of altricial and precocial species. Anim Behav 1972; 20:644-655.
49. Oppenheim RW. Prehatching and hatching behavior: A comparative and physiological consideration. In: Gottlieb G, ed. Behavioral Embryology. New York: Academic Press, 1973:163-244.
50. Casey MB, Sleigh M. Cross-species investigations of prenatal experience, hatching behavior, and postnatal behavioral laterality. Dev Psychobiol 2001; 35:84-91.
51. Kaplan G, Rogers LJ. Birds: Their habits and skills. Sydney: Allen and Unwin, 2001.

CHAPTER 11

A Role of Functional Brain Asymmetry in Human Adaptation

Elena I. Nikolaeva* and Vitaly P. Leutin

Abstract

In the present review the data on distribution of individuals with different sensory and motor asymmetric characteristics are discussed. A joint index that more completely profiles functional sensorimotor asymmetry (i.e., a right- or left-side preference or absence of this preference for the use of hands, feet, eyes and ears) is proposed. It is shown that pronounced variations in asymmetry profiles can be found in different geographical regions. For example, there are a large number (40%) of individuals with the left and symmetrical profiles among the natives of the Far North (the Selkups) compared to the population living in Novosibirsk (19%). The population of Selkups also showed extraordinary low concentrations of cholesterol and cortisol in comparison with Russians. Individuals with the left asymmetry profiles (leading left hand, foot, eye and ear) were found to have a lower incidence of myocardial infarction. Inhalation of a gas mixture to induce hypoxia (10% oxygen and 90% nitrogen) produced more significant increase in the linear blood flow in the left-handed healthy subjects compared to that in subjects with the right profile. It is suggested that in extreme conditions, the best adaptation is characteristic of people with a left asymmetry profile. In this case the adaptation is achieved due to simultaneous activation of both hemispheres. Finally, in individuals with a right asymmetry profile who are under extreme conditions, activation of the left hemisphere is followed by increasing activity of the right hemisphere. We hypothesize that this finding might ultimately result in a failure of the central mechanisms of vegetative regulation.

Introduction

Different findings from studies of functional brain asymmetry in humans have highlighted a variety of possible causes for this phenomenon.[1-5] The most investigated characteristic of lateralized asymmetry is handedness. Although the reported percentage of left-, or right-handed individuals differs from study to study, this could be due to methodological differences between these studies, such as the questionnaires used, the tasks performed, or the action(s) observed.[6-8] Furthermore, handedness may be affected by a number of social and native factors as well as severity of climatic-geographic conditions under which a subject lives.[9] Examination of a variety of asymmetrical sensorimotor characteristics—leading hand, foot, eye and ear— also can result in alternative interpretations of the obtained data. While some authors deny the existence of people with absolute left characteristics (the preference of left side for the use of the hand, foot, eye and ear),[10] others report a high proportion of such individuals in some geographical regions.[11]

*Corresponding Author: Elena I. Nikolaeva—Herzen State University, Brestskii Bulvar 13, apart.14, St. Petersburg, 198328 Russia. Email: klemtina@yandex.ru

Behavioral and Morphological Asymmetries in Vertebrates, edited by Yegor B. Malashichev and A. Wallace Deckel. ©2006 Landes Bioscience.

The main purpose of the present review is to describe the relationships between asymmetry of sensorimotor characteristics and adaptive capabilities of humans. A new concept suggesting an important role of functional sensorimotor asymmetry for human evolution is also proposed.

Determination of Handedness and Functional Sensory and Motor Asymmetries

The definition of handedness in humans is complicated by a variety of terms that have been used by different authors: right-handedness versus left-handedness, dextrality versus sinistrality, ambidexterity versus ambisinistrism, left preference versus right preference, hidden sinistrality versus hidden dextrality, absolute (pure) sinistrality versus absolute (pure) dextrality,[1] and left-hander versus right-hander.[10] Ambidexterity refers to those individuals who are able to use both hands equally well (ambi—two, dextrum—right), whereas ambisinistrism[1] indicates unsuccessful usage of both hands. Mixed-handedness is often used as a term to denote individuals using both hands in turn when performing different tests.[12] There are also such terms as "nonconstant-left-handedness" and "constant-left-handedness".[13] This variety of terms is due both to the variety of methods used for the measurement of handedness and to the complexity of the handedness phenomenon itself.

Handedness initially was characterized by observation of hand usage in different situations and by the use of "preference" questionnaires. These two methods show different results when measuring handedness in the same groups of people.[6] The questionnaire approach, however, has several obvious shortcomings, the most serious of which is reporter bias. Because of this, there has been an increasing awareness by researchers that functional measures of laterality must be used to assess handedness. These studies also underscore the importance of evaluating each lateralized variable not only by administering a questionnaire but also by performance of the task, i.e., on the basis of real action. This may be due to the fact that most people, on self-report questionnaires, do not reliably report their automated actions and have an ideal image of themselves. Therefore, in order to meet the requirements of the society they more frequently self-report themselves as right-handers.[14]

But another question is how many tasks are necessary for correct measurement of handedness. Some authors take just one task, for example hand preference during the use of a machete,[15] while others apply ten tasks or more, for example hand preference during the use of a pen, scissors, needle, toothbrush and so on.[9-11] There is no doubt that the different sets of tasks (or tests) differentially ascribe handedness, because there are many subjects who are either ambidextrous or mixed-handed. Hence the use of only one task may not identify them.[15-17] Rather, previous works using Factor Analysis demonstrated that the concept of handedness is more complex than was initially believed and includes three or six factors.[18-20] Thus it is clear that modern handedness research must include measures that assess a wide variety of variables related to hand movements (three tests or more).

A number of studies completed at the end of the 20th century failed to confirm an association between handedness and certain psychophysiological parameters,[21-23] but rather underscored that there is a special mechanism for supporting the constant quantity of lefthanders.[12,17] The latter findings stimulated investigators not to focus entirely on pure handedness, but to search for other indices of an underlying overall functional laterality, such as the preferential use of feet, eyes and ears. According to this approach, the use of the greatest possible number of sensorimotor variables better describes asymmetry in functional cerebral activity than the use of only one of these parameters.[24]

At present the relationships between the indices of sensory and motor asymmetries are still unclear. A number of authors consider the lateralized preference for the use of the hand, foot, eye and ear as independent variables[6,25] and the combination of states of these parameters may vary during different kinds of activities.[26] Some researchers have suggested that sensory and motor asymmetries of vertebrates evolved as independent variables.[27]

An index consisting of four measurements of laterality (i.e., lateralized preference for the use of hands, feet, eyes and ears that is handedness, footedness, eyedness and eardness) was named the "profile of functional sensorimotor asymmetry".[6,9] Each characteristic (for example, handedness or footedness) was measured by using seven to eight performance tests. If an individual preferred to use only left side in different tests for one characteristic, this characteristic was considered left (for example, left footedness). If an individual preferred to use both the right and left sides in different tests, this characteristic was considered symmetric. Individuals demonstrated right-sided preference for the use of the hands, feet, eyes and ears or a preference for right-side use of at least three of four mentioned characteristics are defined as having a right profile. Similarly, subjects characterized by left-sided preference comparable to those described for the right side are referred to as having the left-side profile. The third group of individuals with symmetric profiles demonstrated complete or partial symmetry of lateral parameters that is they have no any preference in using hands, feet, eyes and ears. The final, fourth, group includes subjects who have mixed combinations of left- or right-side sensorimotor preferences or symmetry for one parameter.

Attempts to Alternate Lateral Preference

Numerous studies have revealed that lateralized motor movements can be enhanced but not reversed by intense training. For example, a coefficient of right-handedness in tennis players has been shown to increase as the time of training is prolonged. This phenomenon is reversible and indeed lateralized hand preference in tennis players decreases when regular exercises are discontinued.[28] Similarly, football players have been reported to use the same leading foot during crucial competitions, suggesting that lateralized foot use increases during games where the pressure to win is increased.[29] These authors have demonstrated that the attempts to teach the sport's beginners to use a nonpreferable foot in complex movements lead to delay of skill mastering.[30]

Unlike not-trained people, the leading leg in skilled fighters can form a new complex with body position that might be controlled by a new brain functional motor system. This gives evidence for a flexibility of the brain, i.e., its ability to create new functional systems for interaction of various muscular groups in the course of training.[19] The right hemisphere of the brain was found to be engaged in the formation of new automatism (for reviews see refs. 6,9,11,21,23).

Thus, many of these observations showed that sensory and motor asymmetry can be enhanced, weakened or combined, but not reversed. The regulation of native asymmetry could lead to delay in sport achievements.[31] Thus the attempts to change the eye and hand preference in baseball players could lead to delay of the results or to disturbances of the binocular vision.[32] The attempts of retraining of left-handed children to write with a right hand lead to neurotics reactions.[33]

The reinforcement of motor asymmetry has been shown to be related to flexible reorganization in the central nervous system. Studies using functional magnetic resonance imaging revealed that the length of the dorsal part of anterior cingulate gyrus of right-handed musicians who start playing musical instruments by the age of seven or eight differs from that of non musicians. This brain region is known to be symmetrically activated in nonmusicians and closely associated with handedness in musicians.[34]

Distribution of the Left-Side and Right-Side Lateralized Behaviors in Different Populations

The distribution of people with different sensorimotor profiles in different populations to a great extent depends on the severity of climatic conditions. According to our data,[6] the majority of professional drivers living in Novosibirsk were characterized with the mixed sensorimotor profiles (52%), 29% of them had the right profiles, 7%—the left and 12%—the symmetric ones. This ratio of people with different profiles is typical for other regions of Russia with temperate climate (for a review see ref. 9).

There is, however, a large number of sinistral individuals in the population of Selkups—the natives of the North of the Tumensk region (Siberia). Only about 41% of the Selkups had the mixed asymmetry profiles, 19% of them showed predomination of the right-side characteristics of functional sensorimotor asymmetry, 10% of them demonstrated the left profile and 30%—the symmetric one.[6] Some other authors also reported a growing number of left-handers or people with symmetric characteristics of the functional profiles among the natives of the North.[35-38] The left-side or symmetric profiles were also found to be most common within the populations living in high mountains.[39-41]

Very interesting data were found in a sample of transit oil workers. They were living in the European part of Russia and working 14 days in a month in North of Siberia, i.e., they have to fly twice a month with time shifts from three to five time zones. Among the workers who worked in such a way during only one year we found the same quantity of left and right profiles as in Novosibirsk population. But among subjects who has a long experience of transit work (during seven years), the frequency of the left-side preference or symmetric profiles was significantly greater.[6,11]

This phenomenon can be explained by a significantly greater incidence of cardiovascular disorders observed in transit workers with predominant right-side functions under sharply changing conditions.[6]

The Selkups—the native inhabitants of North of Siberia—have predominantly left and symmetrical profiles and they also demonstrated lower level of cardiac disorders (the typical disorders of the Russians with right and mixed profiles). A biochemical study has revealed that the population of the Selkups as a whole shows extraordinary low blood concentrations of cholesterol and low density lipoprotein in comparison with the Russians. Both for the Selkups and the Russians the patterns of lipid metabolism were worse for subjects with right and mixed profiles.[42] High rates of right profiles exist among patients with myocardial infarction (Russian people living in Novosibirsk) and these numbers of right profile men become greater as the severity of the cardiac disease increases.[43] Thus it is possible that distributed loading on hemispheres under the extreme adaptation of people with the left and symmetric profiles promotes more effective cardiovascular regulation.

The people with left and symmetrical profiles have the same biochemical reactions and the same probability of myocardial infarction. It is proposed that their hemispheres are more equal than that in the brains of people with right profiles. Thus, in extreme conditions the best adaptation is found in people with symmetrical brain, rather than asymmetrical one.

Peculiarities of Hemispheric Interactions in the Process of Adaptation

Results of several expeditions to Baikal Lake, Primorski Krai, Kamchatka, Pamir and Altai have revealed some commonalities of interhemispheric adaptation as humans have adapted to new environmental conditions. At early stages of adaptation to severe environments (i.e., 2nd or 3rd days of habitation under sub-extreme conditions) the recall of new information or information with an emotional valence was shown to increase, while recall of neutral information worsened.[8] This finding implies that an active selection of new, unusual and highly-significant stimuli by the language dominant left-hemisphere might be the first psychophysiological reaction under extreme conditions. These phenomena were observed both during processing of verbal and nonverbal (tones) stimuli.[6,44]

Traditional experimental methods for the study of hemispheric asymmetry, like dichotic listening, revealed a dominant role of the right ear for processing of neutral words.[45] Our experiments found that more effective retrieval of novel words in adaptation occurs due to better recall of the words presented to the right ear (processed by the left hemisphere).[6,44] In addition, retrieval of emotional words was enhanced due to better recall of the words presented to the left ear (processed by the right hemisphere). It is conceivable that processing of verbal information under new conditions is associated with selection of unusual and emotional types of information and as a result, activation of the left hemisphere is followed by increasing activity of the right one. It is also possible that the abilities of both cerebral hemispheres are required for the process of adaptation.[6,44] The right hemisphere is subjected to feedback control to a lesser degree than the left one.[46] Having greater autonomy, the right hemisphere is selectively involved in the assessment of uncertain environmental cues and the prediction of improbable events.[47] At the same time the constant monitoring of the environment and the determining of the importance of various different actions under usual conditions is best provided by the left hemisphere.[48]

An urgent mobilization of the left hemisphere resulting from sharp shifts of environmental variables may occur during adaptation. A certain level of this dissociation seems to affect the interaction between the hemispheres due to the redistribution of their activities through the corpus callosum. We have shown before that exposure to novel environmental conditions facilitates cross-transfer of information between the left and right hemispheres.[44] Thus, formation of a new behavioral stereotype more adequate to changing conditions becomes necessary. The prevalence of the right hemisphere for creation of the integral image of the environment makes this possible.

Activation of the right hemisphere during adaptation to novel environmental conditions has been shown using various experimental paradigms.[6,49-50]

Exposures to novel stimuli or situations that provoke emotional responding and involve the attachment of an emotional valence to the stimuli are tasks well-handled by the right hemisphere. At the same time, the left hemisphere, which is superior for analytical functions, is integrating the new information.

In summary, we believe that adaptation leads to asymmetrical activation of the cerebral hemispheres. Unusual and new signals are primary processed by the left hemisphere. Conversely, information that has a high emotional valence may selectively engage the right hemisphere. And finally, an initial superiority of the left hemisphere is established. Thus interaction between the two hemispheres is facilitated by exposure to unusual and emotionally important conditions and by extreme conditions that require a high level of adaptation in order to survive.

Comparison of Adaptive Mechanisms in Humans with the Different Profiles

Using the method of transcranial Doppler sonography, the rate of linear blood flow in the cerebral hemispheres of subjects with left or right profiles has been analyzed during exposure to oxygen deficiency. We have also examined whether or not similar adaptive changes in brain regulatory mechanisms occurred in populations living under conditions of hypoxia, for example, hypoxic hypoxia in the mountain regions, or hypoxia due to extreme cold in the Far North.[11,51]

Our study showed that before the inhalation of a gas mixture the rate of the blood flow was greater in the left hemisphere than in the right one in those subjects who had the right profile of sensorimotor asymmetry. Activation of the hemispheres was also shown to be accompanied by an enhancement of metabolic processes. The values of the blood flow rate in subjects with left or symmetric profiles were approximately the same for the left and right hemispheres indicating the involvement of both hemispheres in the regulation of metabolic processes.

Inhalation of a gas mixture to induce hypoxia (10% oxygen and 90% nitrogen) produced an increase in the linear blood flow in both hemispheres in all groups of subjects. However, the observed changes were more pronounced in the hemispheres of the subjects with left profiles. It should be noted that after the prolonged exposure to hypoxia the blood flow rate was significantly reduced in the left hemisphere of the subjects with the right asymmetry profiles compared to the group with the left asymmetry profiles.[51]

There is much evidence in the literature that the cerebral hemispheres are differentially involved in the control of vegetative functions. Numerous data indicate a specialized role for the right hemisphere in the regulation of endocrine functions and immunity.[52,53] These functions are also known to be closely connected with subcortical brain structures.[54] Greater blood flow rate in the right hemisphere in subjects with left profiles suggests a more effective regulation of metabolic processes by their right hemisphere compared to that in subjects with right profiles. Therefore, in extreme conditions the activation of the right hemisphere in right profile subjects will cause in decrease of the blood flow in both hemispheres, whereas in subjects with left profiles blood flow in both hemispheres will remain normal. This means better adaptation of the subjects with left profiles.

During extreme oxygen deprivation people with right profiles showed initial activation of the left hemisphere followed, later, by the activation of the right hemisphere.[51] Subjects with left or symmetric profiles activated both hemispheres simultaneously. In these subjects the loading is distributed between both hemispheres. This may, during extreme adaptation, prevent failure of the central mechanisms of vegetative regulation (and as a consequence, cardiac disorders).

There are two explanations of consecutive changing of hemispheric activity in people with right profiles. As functional brain asymmetry, probably, reflects dominant-subdominant relationships in the brain,[55] it is possible that activation of the right hemisphere and its emotion-related limbic structures is the result of hyperactivation of the left hemisphere in the first period of adaptation. In this case the activation of the right hemisphere could occur on a background of decreasing of a blood circulation and dysfunction in the left one. The other explanation is that the left hemisphere may transfer active functions to the right hemisphere because the right hemisphere is responsible for creation of new programs of interaction with the changed environment.[23,24] From our point of view, participation of both of these mechanisms is possible during normal functioning of the integrated cerebral hemispheres. However, activation of the right hemisphere can simultaneously worsen the regulation of blood circulation in the whole organism.

Thus, in extreme conditions the best adaptation is found in people with less lateral brain specialization (left and symmetric profiles). In this case the adaptation is achieved due to simultaneous work of both hemispheres. The people with right asymmetry profiles are characterized by other type of adaptation—the sequence of reciprocal relationships with one activated hemisphere and a suppressed other one.

In many languages the word "left" means not only "direction" but also "bad" and the word "right" has an additional meaning "good" (for a review see ref. 24). Unfortunately, in many previous studies researchers tried to show the connection of the left handedness exclusively with pathology[56] and even with homicide.[15] However, a more careful examination shows that left-handedness only occasionally results from a pathological process. The majority of left-handers have no clearly definable pathology.[57] It is possible that lateralized hand usage is due to hemispheric interconnections in people with different sensorimotor profiles. This peculiarity leads to a better adaptation of people with right profiles in temperate climate and with left profiles in a severe climate.

References

1. Luria AR. Human brain and psychic processes. Moscow: Moscow University Press, 1963, (In Russian).
2. Bradshaw JL. The evolution of human lateral asymmetries: New evidence and second thoughts. J Hum Evol 1988; 17:615-637.
3. Galaburda AM, LeMay M, Kemper TL et al. Right-left asymmetries in the brain. Science 1978; 199(4331):852-856.
4. Hellige JB. Hemispheric Asymmetry. Cambridge: Harvard University Press, 1993.
5. Sperry RW. Some effects of disconnecting the cerebral hemispheres. Bioscience Reports 1982; 2(5):265-276.
6. Leutin VP, Nikolaeva EI. Psychophysiological mechanisms of adaptation and functional brain asymmetry. Novosibirsk: Nauka, 1988, (In Russian).
7. Crovitz HF, Zener KA. A group-test assessing hand and eye dominance. Am J Psychol 1962; 75:271-276.

8. Barnsley RH, Rabinovitch MS. Handedness: Proficiency versus stated preference. Percep Mot Skills 1970; 30:343-362.
9. Bragina NN, Dobrochotova TA. Functional Human Asymmetries. 2nd ed. Moscow: Medicina, 1988, (In Russian).
10. Dobrochotova TA, Bragina NN. Lefthanders. Moscow: Kniga, 1994, (In Russian).
11. Leutin VP, Nikolaeva EI. The myths and reality of functional brain asymmetry. St. Petersburg: Retch, 2005, (In Russian).
12. Annett M. Left, right, hand and brain: The right shift theory. London: Lowrence Erlbaum, 1985.
13. Peters M. Subclassification of nonpathological left-handers poses problems for theories of handedness. Neuropschologia 1990; 28:282-286.
14. Bryden MP. Measuring handedness with questionnaires. Neuropsychologia 1977; 15:617-624.
15. Faurie C, Raymond M. Handedness, homicide and negative frequency-dependent selection. Proc R Soc Lond B 2004; 271:43-45.
16. Alonso SJ, Navarro E, Santana C et al. Motor lateralization, behavioral despair and dopaminergic brain asymmetry after prenatal stress. Pharmacol Biochem Behav 1997; 58(2):443-448.
17. Annett M. Predicting combinations of left and right asymmetries. Cortex 2000; 36(4):485-505.
18. Healey JM, Liederman J, Geschwind N. Handedness is not a unidimensional trait. Cortex 1986; 22:33-41.
19. Nikolaenko NN, Afanasjev CV, Micheev MM. Organization of motor control and peculiarities of functional brain asymmetry of fighters. Physiology of Human 2001; 27(2):68-75, (In Russian).
20. Steenhuis RE, Bryden MP. Different dimensions of hand preference that relate to skilled and unskilled activities. Cortex 1989; 25:289-297.
21. Bryden MP. Laterality. Functional Asymmetry in the Intact Brain. New York, London: Academic Press, 1982.
22. Annet M, Manning M. Reading and a balanced polymorphism for laterality and ability. J Child Psychol Psychiat 1990; 31:511-529.
23. Hellige JB. Hemispheric Asymmetry: What's Right and What's Left? Cambridge: Harvard University Press, 2001.
24. Springer S, Deutsch G. Left Brain Right Brain: Perspectives From Cognitive Neuroscience. 5th ed. New York: WH Freeman and Company, 1998.
25. Mosidze VM, Riginashvili PC, Samadishvili ZV et al. Functional brain asymmetry. Tbilisi: Mezniereba, 1977, (In Russian).
26. Treffner PJ, Turvey MT. Symmetry, broken symmetry and handedness in bimanual coordination dynamics. Exp Brain Res 1996; 107:463-478.
27. Malashichev YB, Wassersug RJ. Left and right in the amphibian world: Which way to develop and where to turn? Bioessays 2004; 26:512-522.
28. Iljyn EP. Differential Psychophysiology. St. Petersburg: Piter, 2001, (In Russian).
29. Mednikov PN. To method of technical training of high skilled footballers: Method Letter. Minsk: Respub. Sci-Method Library on Physical Culture, 1975, (In Russian).
30. Lebedev VM, Mednikov RN. Right-left. Sport games 1977; 8:7-9, (in Russian).
31. Karjagina NV. Lateral Limitation of the Loading in the Process of Sportsmen Training. Krasnodar: Krasnodar Pedagigical University Press, 1996, (in Russian).
32. Portal JM, Romano PE. Major review: Ocular sighting dominance in a collegiate baseball team. Binocul Vis Strabismus Quart 1998; 13(2):125-132.
33. Semenovich AV. Interhemispheric Organization of Psychic Processes in Lefthanders. Moscow: Moscow State University Press, 1991.
34. Amunts K, Schmidt PF, Schleicher A et al. Postnatal development of interhemispheric asymmetry in the cytoarchitecture of human area 4. Anat Embryol Berl 1997; 196(5):393-402.
35. Arshavsky VV. Interhemisphere asymmetry in the system of searching activity (to the problem of human adaptation in polar regions of North East of USSR). Vladivostok: Acad Sci USSR Press, 1988, (In Russian).
36. Hasnulin VI. Disadaptation, pathology and brain asymmetry. Arhiv Psihiatrii 1997; 12-13:23-26.
37. Rotenberg VS, Arshavsky VV. Psychophysiology of hemispheric asymmetry: The "Entropy" of right hemisphere activity. Integr Physiol Behav Sci 1991; 26:183-188.
38. Rotenberg VS, Arshavsky VV. Right and left brain hemispheres activation in the representatives of two different cultures. Homeostasis 1997; 38(2):49-57.
39. Leutin VP, Chuhrova MG, Krivoshekov SG. The connection of alcogolism with peculiarities of functional asymmetry of Tuvinians. Physiology of Human 1999; 25(2):67-70, (In Russian).
40. Ijikova EA. Psychophysiological and Morphofunctional Characteristics of the Russian and Altaian Adolesents 14-15 Years. Novosibirsk: Novosibirsk Pedagogic University Press, 2000, (In Russian).

41. Tinalieva BK. Interhemispheric brain asymmetry of mountain's habitants of Kirgizstan. In: Fokin VF, ed. Actual Problems of Functional Brain Asymmetry. Moscow: Isd-vo NII golovnogo mosga RAMN, 2003: 310-315, (In Russian).
42. Nikolaeva EI, Oteva EA, Leutin VP et al. Relationships between left hemisphere predominance and disturbances of lipid metabolism in different ethnic groups. Int J Cardiol 1995; 52(3):207-211.
43. Nikolaeva EI, Oteva EA, Nikolaeva AA et al. Prognosis of myocardial infarction and brain functional asymmetry. Int J Cardiol 1993; 42:245-248.
44. Leutin VP. Adaptation dominate and functional brain asymmetry. Vestnik of Russian Academy of Medical Sciences 1998; 10:10-13, (In Russian).
45. Kimura D. Cerebral dominance and the perception of verbal stimuli. Canad J Psychol 1961; 15:166-171.
46. Kostandov EA. Psychophysiology of Consciousness and Unconsciousness. St. Petersburg: Piter, 2004, (In Russian).
47. Meerson YAA. About the role of left and right brain hemispheres in the processes of probability prognosis. Physiology of Human 1986; 12(5):723-731, (In Russian).
48. Goldberg E, Costa LD. Hemisphere differences systems. Brain and Lang 1981; 14:144-173.
49. Volf NV. Dynamics of concurrent interconnection of verbal and manual activity in adaptation and readaptation after transmeridian flight. Physiol of Human 1991; 17(6):142-146, (In Russian).
50. Fokin VF, Ponomareva NV. Dynamic characteristics of functional interhemispheric asymmetry. Moscow: Scientific World, 2004:349-369, (In Russian).
51. Leutin VP, Platonov YAG, Divert GM et al. The role of brain asymmetry in subjective estimation formation of the state during interruptive hypoxia training. Human Physiology 2002; 28(1):67-70, (In Russian).
52. Abramov VV, Abramova TYA. Asymmetry of Nervous, Endocrine and Immune Systems. Novosibirsk: Nauka, 1996, (In Russian).
53. Gerendai I, Halasz B. Neuroendocrine asymmetry. Frontal Neuroendocrinology 1997; 18:354-381.
54. Kamenskaja VM, Bragina NN, Dobrohotova TA. To the question about functional connections of right and left brain hemispheres with different parts of median structures of right handers. Proceedings of Moscow Sci. Inst of Psychiatry MH RF. Moscow: Medicina, 1976, (In Russian).
55. Bianki V. The Mechanism of Brain Lateralization. New York: Gordon and Breach, 1993.
56. Geschwind N, Galaburda AM. Cerebral Lateraization: Biological Mechanism, Associations and Pathology. Cambridge: MII Press, 1987.
57. Coren S. The Left-Hander Syndrome—The Causes and Consequences of Left-Handedness. New York: Free Press, 1992.

CHAPTER 12

Functional Asymmetry in Hematopoietic, Immune and Nervous Systems

Valery V. Abramov,* Irina A. Gontova and Vladimir A. Kozlov

Abstract

We report a series of three experiments that suggest that hemispheric dominance for paw preference is related to asymmetries in peripheral physiology. First, we report that bone marrow cells taken from the left femoral bone of (CBA × C57Bl/6) F_1 mice are functionally more active in left-pawed recipients than in right-pawed ones. Second, we found that the brain asymmetrically interacts with the thymus lobes in the regulation of the humoral immune response. Finally, the intensity of Delayed-Type Hypersensitivity (DTH) inflammatory reaction in (CBA × C57Bl/6) F_1 mice depends on the functional asymmetry of regional lymph nodes and paw preference. These findings are discussed in the setting of the integrated Hematopoietic, IMmune, Endocrine and Nervous systems, i.e., the HIMEN system.

Introduction

Previous studies of asymmetry in the cerebral hemispheres revealed structural, sensory, motor and molecular differences in the brain.[1-11] These asymmetries appear to regulate a wide array of biological processes important to normal functioning of the nervous system, including ontogeny, sexual dimorphism, intellectual abilities, and adaptation to extreme factors.[1-11] More recently, it was reported that the cerebral hemispheres asymmetrically contribute to the regulation of the hematopoietic, immune and endocrine systems (HIMEN system).[12-22] This paper reviews the role of asymmetry in the HIMEN systems and, additionally, reports new data from our laboratory on this topic.[23,24]

Materials and Methods

Two-month old male (CBA × C57Bl/6) F_1 mice were obtained from the Tomsk nursery and kept in a vivarium in plastic cages, 10 animals in each. Mice received *ad libitum* food and water. Motor asymmetry was assessed by paw preference in taking food, as described previously.[25] The test was repeated three times with a three-day interval. From this mice were divided into three groups: left-handed, right-handed and ambidextrous. Ambidextrous mice were not used in the experiments. For the purposes of this manuscript, we assume that preferred paw use is under the control of the contralateral cerebral hemisphere and reflects hemispheric dominance for motor activity. Thus we characterized the subjects

*Corresponding Author: Valery V. Abramov—Research Institute of Clinical Immunology of SB RAMS, 14 Yadrintsevskaya ulitsa, Novosibirsk, 630099, Russia.
Email: valery_abramov@mail.ru

Behavioral and Morphological Asymmetries in Vertebrates, edited by Yegor B. Malashichev and A. Wallace Deckel. ©2006 Landes Bioscience.

into either "left-pawed/right hemisphere dominant" or "right-pawed/left hemisphere dominant" groups.

Design for Experiments 1-3

Experiment 1

Experiment 1 investigated the relationship between paw preference and exogenous colony formation (i.e., the number of eight-day colony-forming units in spleen – CFUs-8) following irradiation treatment to destroy the hematopoietic function of the bone marrow.[26]

The recipient mice were prepared by exposure to a whole-body-radiation dose (950 R).[26] Marrow cells were obtained from the femora of mice-donors, suspended in saline, and the nucleated cells were counted in a hemacytometer. These suspensions were kept at ice-water temperature until they were injected. The recipient animals then received an intravenous injection with 10^5 of marrow cells (0.5 ml). The effect of functional asymmetry of the bone marrow on hematopoiesis was evaluated by injections of cells from the left or right femoral bone of left- or right-pawed donors to irradiated recipient animals (left- or right-pawed).[22] Eight days after injection of the transplanted cells, the mice were killed and the number of colonies in their spleens was determined.[26]

Experiment 2

For Experiment 2, the mice recipients were thymectomized to evaluate the role of the cells from either left or right thymus lobes in the humoral immune response (IR). Five weeks after thymectomy, left and right-pawed recipient animals were intravenously injected with thymocytes from the right or left thymus lobes of left-pawed syngeneic donors (10^7 cells/mouse). These recipient mice were subsequently immunized with sheep erythrocytes (SE) 10 days after thymocyte administration.[20,23] Antibody producing cells (APC) in the spleen were counted by the method of Cunningham four days after immunization.

Experiment 3

Experiment 3 examined the definition of cellular immune response (DTH: delayed-type hypersensitivity) in the hind and fore paws in left- and right-dominant hemisphere mice, with each animal studied twice within a four-month interval.

Mice-recipients were thymectomized under either anesthesia. Then, five weeks after the operation, mice were intravenously injected with cells from either left or right lobe of thymus (10^7 cells/mouse) from right-dominant or left-dominant mice-donors. Ten days later mice-recipients were intraperitoneally immunized with 0.5 ml of 5% SE. On the fourth day of the Delayed-Type Hypersensitivity (DTH) reaction was used as an in vivo measure of antigen specific T-lymphocyte reactivity after Yoshikai et al method.[27] Delayed-Type Hypersensitivity reaction was estimated in the left hind paw. In order to study the cellular immune response, 50% suspension of SE in 0.05 ml of physiologic salt solution was injected under aponeurosis of the left fore-paw. 0.05 ml of physiologic salt solution was injected under aponeurosis of the right fore-paw as a control. The right hind paw was injected with the same quantify of physiologic salt solution as a control. The cross-section of a paw at the site of injection was measured 24 hours later.[24] The index of reaction (IR) was calculated for each mouse with the formula: $IR = (P_o - P_c)/Pc$, where P_c-paw cross-section in control mice and P_o-paw cross-section in experimental mice.

Statistical Analysis

The results were statistically processed using ANOVA with least significant differences procedures (experiment 1), Student's t-test for independent samples (experiment 2) and Student's t-test for independent samples or Mann-Whitney U-test (experiment 3).

Results

Experiment 1

Evaluation of functional asymmetry of the bone marrow showed significant differences in the formation of CFUs-8 only in irradiated left-pawed recipients receiving bone marrow cells from the right or left femoral bones of left-pawed donors (P = 0.021; Fig. 1A). No appreciable differences in the formation of CFUs-8 were observed in cases when left and right femur marrow cells were transplanted from right-pawed donors to left-pawed recipients, or marrow cells from either femur of left- and right-pawed donors were injected to right-pawed recipients. The results indicate that the hematopoietic potential of the bone marrow from the right and left femoral bones is different, in agreement with previous reports.[16-19] In addition, manifestation of asymmetry of bone marrow hematopoietic functions depends on motor asymmetry of donors and recipients.

Irradiated recipients (either left or right-handed) transplanted with cells from the right femoral bone of right-pawed donors had greater hematopoiesis in the spleens of recipients compared to cells from left-pawed donors (correspondingly, P = 0.014 and P = 0.000013; Fig. 1B). No significant differences between the groups were observed after transplantation of the bone marrow from the left femoral bone from either left- or right-pawed donors. Hence, the bone marrow from the right femoral bone of right-pawed donors possesses higher hematopoietic potential compared to that of left-pawed animals. In addition, hematopoietic functions of the bone marrow from the left femoral bone did not depend on the motor asymmetry of the donors.

Analysis of the relationship between motor asymmetry in irradiated recipients and the number of CFUs-8 showed that the differences between the groups were significant only when irradiated mice were injected with bone marrow cells from the left femoral bone of left-pawed donors: the number of colonies in the spleens of left-pawed recipients was significantly higher than in right-pawed animals (P = 0.003; Fig. 1C). On the other hand, the formation of CFUs-8 in right- and left-pawed recipients did not differ significantly in all other combinations. These findings indicate that bone marrow taken from the left femoral bone is functionally more active in left-pawed recipients than in right-pawed ones. Conversely, when taken from the right femoral bone, hematopoietic functions of the bone marrow do not depend on the recipient's motor asymmetry.

Our findings suggest that the formation of CFUs-8 in recipients depends on the motor asymmetry of donors and recipients and on the source (right or left femoral bone) of the bone marrow. First, bone marrow cells from the left femoral bone can induce the formation of a higher number of colonies compared to those from the right femur. Second, bone marrow cells from right-pawed donors ensure more intensive hematopoiesis compared to those from left-pawed donors. Finally, in left-pawed recipients the number of splenic colonies was higher than in right-pawed ones.

It is noteworthy that the differences between the groups were also significant in cases when two of the three studied parameters were considered in different comparisons. For example, hematopoiesis depends on both bone marrow asymmetry and motor asymmetry of cell donors. Injection of bone marrow cells from the right femoral bone of right-pawed donors to right-pawed recipients led to the formation of significantly more colonies than after injection of cells from the left femoral bone of left-pawed donors to the same recipients (P = 0.00047). The number of CFUs-8 differed significantly in left-pawed recipients transplanted cells from the right femoral bone of left-pawed donors or cells from the left femoral bone of right-pawed donors.

Statistically significant between-group differences also were found when bone marrow asymmetry and motor asymmetry of recipients were considered in different comparisons. When left bone marrow cells from right-pawed donors were injected to left-pawed recipients, there were greater numbers of splenic colonies than when right bone marrow cells from left-pawed donors

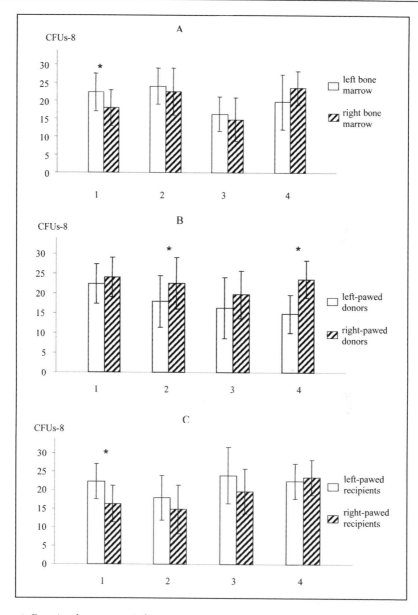

Figure 1. Functional asymmetry in bone marrow (A), and hemispheric asymmetry of donors (B) and recipients (C) during formation of CFUs-8 (the number of eight-day colony forming units in spleen) in (CBA × C57BL/6) F1 mice. A) 1: left-pawed recipients, left-pawed donors; 2: left-pawed recipients, right-pawed donors; 3: right-pawed recipients, left-pawed donors; 4: right-pawed recipients, right-pawed donors. B) 1: left-pawed recipients, left bone marrow; 2: left-pawed recipients, right bone marrow; 3: right-pawed recipients, left bone marrow; 4: right-pawed recipients, right bone marrow. C) 1: left-pawed donors, left bone marrow; 2: left-pawed donors, right bone marrow; 3: right-pawed donors, left bone marrow; 4: right-pawed donors, right bone marrow. Values represent the mean ± SD for N = 20 in each group. The data are analyzed by ANOVA. *: $P < 0.05$.

were injected into left-pawed recipients (P = 0.0013). In other words, the combination of donor bone marrow asymmetry and recipient motor asymmetry is significant for the degree of hematopoiesis.

Interactions between bone marrow donor and recipient motor asymmetry also resulted in significant changes between the groups. Such differences were observed when, for example, recipients were transplanted with bone marrow cells from the right femoral bone, but motor asymmetry in the groups was matched in both donors and recipients. Between group differences were also significant when right-pawed recipients were transplanted with right bone marrow cells from left-pawed donors, and when the left-pawed recipients were transplanted with right bone marrow cells from right-pawed donors (P = 0.00008). With bone marrow cells from the left femoral bone, the differences between the groups were significant only when the donor-recipient pair in each group was reversed for motor asymmetry (P = 0.0016). These data indicate combined effects of motor asymmetry of bone marrow donors and recipients on the number of splenic colonies.

Thus, our experiments once more confirm the existence of functional asymmetry of the bone marrow and, for the first time, demonstrate that the capacity of (CBA ξ C57Bl/6) F_1 donor bone marrow cells to generate a hematopoietic response depends on motor asymmetry of both the donors and recipients and on the functional asymmetry of the bone marrow.[16-19,22]

Experiment 2

Previous work indicated that immunological parameters and the incidence and severity of autoimmune diseases in humans differ between left- and right-handers.[14] However, the mechanisms for these differences remains unknown.

Previously, we hypothesized that the relationship between the cerebral hemispheres and thymus lobes underlie the differences in humoral IR. Earlier work in our laboratory demonstrated functional asymmetry of the immune organs; e.g., thymus, bone marrow, and lymph nodes.[16-19] To evaluate the role of functional asymmetry in the nervous and immune systems, thymectomized right- and left-pawed recipient mice received thymocytes from the left or right thymus lobe of left-pawed donors.

The results of two repeated experiments (Fig. 2A) show the difference in the humoral immune response in mice that previously received intravenous injection of cells from the right and left lobes of the thymus. In both experiments the recipients with left-hemisphere dominance that received an injection of cells from the left lobe of the thymus demonstrated a significantly higher immune response to SE than the left-dominant hemisphere mice that received an injection of cells from the right lobe of the thymus (P < 0.05). In right-dominant hemisphere mice, significant changes in the immune response were only evident in the second experiment. Thus it is unclear if the differences obtained from right-dominant hemisphere mice represent a real effect.

Figure 2B shows the interaction between motor asymmetry and the side of the thymus lobe on IR responses. Mice with left hemisphere dominance receiving cell injection from the left lobe of the thymus had a significantly higher IR to SE than the mice with right hemisphere dominance (P < 0.05). Conversely, IR was not affected by transplants from the right lobe of the thymus.

Figure 2C shows the difference in IR to SE in the right-dominant and left-dominant hemisphere mice receiving thymocyte injections from the ipsi- or contralateral lobes of the thymus with regard to the dominant brain hemisphere. Only in the second experiment was IR in left-dominant hemisphere mice receiving cell injections from the left lobe of thymus significantly higher than in the right-dominant hemisphere mice receiving cell injections from the right lobe of the thymus (P < 0.05). In mice receiving cell injections from the lobe contralateral to the dominant hemisphere lobes of the thymus, there were no differences in IR between the left-dominant and the right-dominant hemisphere mice.

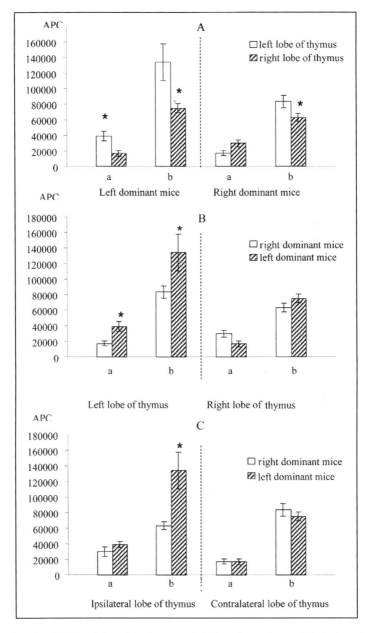

Figure 2. The role of different lobes of the thymus and different hemispheres of the brain to the number of antibody-producing cells (APC) in the spleens of mice. A) APC number in spleens of left-dominant and right-dominant hemisphere recipients. B) Number of APC in the spleens of mice receiving cells from the left and right lobe of the thymus. C) The number of APC in the spleens of mice receiving cells from the ipsilateral or contralateral lobe of the thymus according to the dominant hemisphere of the recipients. On each graph—a: data from the first experiment; b: data from the second experiment. Values at this figure show the mean ± SEM, N = 10 in each group. The data are analyzed with Student's t-test for independent samples. *: significant differences at $P < 0.05$. Reproduced from reference 23, with permission from S. Karger AG.

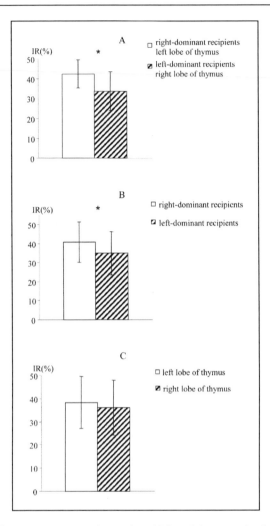

Figure 3. The role of motor asymmetry and contralateral lobes of thymus in the development of DTH reaction in (CBA × C57Bl/6) F1 mice. A) DTH reaction was compared in right-dominant recipients that received thymocytes of the left lobe of right-dominant donors, with DTH reaction in left-dominant recipients that received thymocytes of the right lobe of right-dominant donors. B) DTH reaction was compared in right-dominant and left-dominant recipients. C) DTH reaction was compared in mice that received thymocytes of left and right lobes of right-dominant donors. IR: index of reaction. Value represent the mean ± SD, N = 20 in each group. The results are analyzed with Student's *t*-test for independent samples. *: significant difference at $P < 0.05$. Reproduced from reference 24, with permission from S. Karger AG.

Thus, our experiments once more confirmed the existence of functional asymmetry of the thymus[16-19] and for the first time demonstrate that asymmetry in the nervous and immune systems plays an important role in the development of humoral IR.[20,21,23]

Experiment 3

In order to test the influence of asymmetry of the thymus on cellular immune response, mice were thymectomized and then injected with cells from the contralateral thymus (Fig. 3). The DTH reaction in the back left paw was then studied as a function of left versus

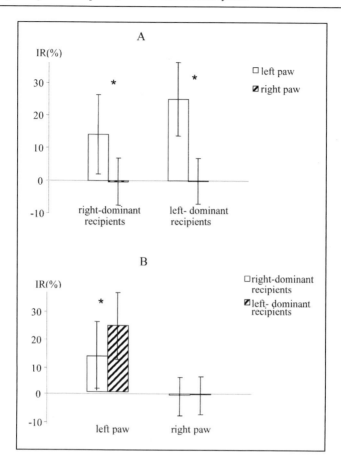

Figure 4. The role of motor asymmetry of brain hemispheres and functional asymmetry of regional lymph nodes in the development of DTH reaction in fore paws of (CBA × C57Bl/6) F1 mice. A) DTH value was compared left and right paws of right-dominant mice and in left and right paws left-dominant mice. B) DTH value was compared in the left paw of right and left-dominant mice and in the right paw of right and left-dominant mice. IR: index of reaction. Value represent the mean ± SD for N = 35/group. The results were analyzed with Mann-Whitney U-test. *: significant difference at $P < 0.05$. Reproduced from reference 24, with permission from S. Karger AG.

right-dominant hemisphere donors. The injection of thymocytes from right-dominant hemisphere donors resulted in significant differences in DTH reaction between left- and right-dominant hemisphere recipients ($P < 0.05$). At the same time, our experiments failed to discover any pronounced role of thymus asymmetry in the formation of DTH reaction.

In order to test the influence of asymmetry of regional lymph nodes on the regulation of cellular immune response, we compared the DTH reaction in left and right paws of mice. We found that the extent of the DTH reaction to SE in the front paws of (CBA × C57Bl/6) F_1 mice depends not only on whether the antigen is injected into the left or right paw but also on the motor asymmetry (Fig. 4). While comparing the DTH reaction in the left and right hind paws of mice we showed that in both it was more pronounced in the left paw than in the right one (Fig. 5; $P < 0.05$). The data obtained continue to support the hypothesis that functional asymmetry exists within bilateral lymph nodes located near the forming cellular immune reaction. Hence, the results obtained show that the degree of DTH reaction in (CBA × C57Bl/6)

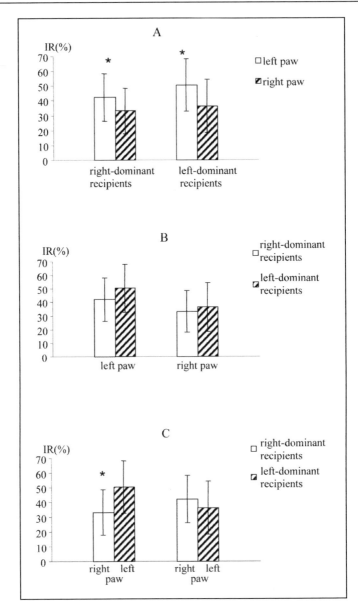

Figure 5. The role of motor asymmetry of brain hemispheres and functional asymmetry of regional lymph nodes in the development of DTH in hind paws of (CBA × C57Bl/6) F1 mice. A) DTH value was compared in left and right paws of right-dominant mice and in left and right paws of left-dominant mice. B) DTH value was compared in the left paw of right- and left-dominant mice and in the right paw of right and left-dominant mice. C) DTH reaction was compared in the right paw of right-dominant and in the left paw of left paw of left-dominant mice; in the left paw of right-dominant and in the right paw of left-dominant mice. IR: index of reaction. Value represent the mean ± SD, N = 30 in each group. The results were analyzed with Student's *t*-test for independent samples. *: significant difference at *P* < 0.05. Reproduced from reference 24, with permission from S. Karger AG.

F_1 mice depends on the functional asymmetry of regional lymph nodes and paw preference. The thymus functional asymmetry is not of importance in generation of the DTH reaction.[24]

Discussion

The asymmetry of the brain hemispheres has been studied for a long time. While studying this phenomenon, numerous facts on structural, functional and molecular-biological differences of the hemispheres have been accumulated.[1-11] Based on these data, some studies have been conducted showing that there is asymmetry of not only the brain hemispheres but also of the neuroendocrine system as a whole, including the gonads, adrenal glands and lobes of the thyroid gland.[12,13,16]

There are publications on the differing roles of the brain hemispheres in the regulation of the immune response.[14,15,18]

At the same time, it is known that the hematopoietic and lymphoid organs (bone marrow, thymus, lymph nodes, etc.) as well as the brain hemispheres are presented by two morphologically divided lobes. This enabled us to suppose and prove that not only the neuroendocrine system but also the hematopoietic and immune systems demonstrate the functional asymmetry of bilateral organs.[16-24] So, our experiments once more confirm the existence of functional asymmetry of the bone marrow and for the first time demonstrate that the capacity of (CBA × C57Bl/6) F_1 donor bone marrow cells to generate a hematopoietic response depends on motor asymmetry of these cells' donors and recipients and on the functional asymmetry of the bone marrow.[16-19,22] Besides, our experiments once more confirmed the existence of functional asymmetry of the thymus[16-19] and for the first time demonstrate that asymmetry in the nervous and immune systems plays an important role in the development of humoral IR.[20,21,23] At last, the results obtained show that the degree of DTH reaction in (CBA × C57Bl/6) F_1 mice depends on the functional asymmetry of regional lymph nodes and paw preference. The thymus functional asymmetry is of insignificant importance in DTH reaction.[24]

It is known that autonomic nerves are well presented in the bone marrows, thymus lobes and lymph nodes of mice where, together with some cells, they form the neuroendocrine environment that influences the maturation of cells.[28,29] These data together with our results on differences in the functional properties of cells from contralateral bone marrow and lymph nodes, and thymocytes from the thymus lobes allow us to suggest the following: (1) there are differences in the sympathetic and parasympathetic innervation of the contralateral lobes of the organs; (2) differences in the functional properties of cells from the contralateral lobes of the organs are caused by differences in the neuroendocrine environment of the lobes, i.e., preferential influence of catecholamines, acetylcholine and peptides on the cells.

It has been established that sympathetic and parasympathetic activity is preferably regulated by different brain hemispheres.[30-32] Sympathetic activity is preferentially regulated by the right hemisphere, whereas parasympathetic activity is regulated by the left hemisphere. In this connection, one can speak about the lateralizing effect of the brain hemispheres on the organs by creating differences in the neuroendocrine environment on the right and left sides.

At the same time, our data, for example, on the different roles of cells from the contralateral lobes of the thymus in the formation of a humoral immune response confirm the supposition that thymocytes from the mentioned lobes have different functional properties. Moreover, we speculate that the number of T-helper precursors of type 2 might be different in the thymus lobes and/or their precursors are at different stages of differentiation. Since the injection of thymocytes from the left thymus to thymectomized recipients is accompanied by the greatest effect on the formation of a humoral immune response, there might be more T-helper precursors in the left lobe and/or they are more mature in comparison with cells from the right lobe. That sympathetic and parasympathetic activity is mainly regulated from different brain hemispheres can help to explain their role in the formation of a humoral immune response in thymocyte recipients. If functional differences in cells from the contralateral lobes of the thymus of donors define the preferential influence of catecholamines and acetylcholine, this can also explain the role of the hemispheres in the formation of antibody-forming cells in the recipients who received thymocytes from a given lobe. For example, specific receptors to definite neuromediators might be expressed on the surface of the thymocytes.

We speculate that the more pronounced DTH reaction on the left than on the right may be connected to asymmetry of peripheral innervation of contralateral lymph nodes that, in its turn, is controlled by brain hemispheres. That is, peripheral vegetative innervation of regional lymph nodes, regulated by brain hemispheres, might be very important in asymmetrical development of reactions of cellular immunity in fore- and hind limbs of mice. The data, obtained by Tarkowski and co-authors,[33,34] testify an important role of sympathetic innervation in DTH reaction. So, they showed that stroke lateralized T-cell-mediated cutaneous inflammation. This effect may be mediated by alteration of the cutaneous sympathetic nerve traffic.[33] This group has also demonstrated lateralization of cutaneous inflammatory responses in patients with paresis after poliomyelitis. This lateralization of DTH responses is related to deficiencies in motor and sympathetic innervation of the paretic extremity.[34]

It is possible, that sympathetic and parasympathetic activity, which is mainly regulated by different brain hemispheres, can help to explain their role in the formation of CFUs-8 in mice recipients.

Thus, our data allow us to speak about the asymmetry of the integrated Hematopoietic, IMmune, Endocrine and Nervous systems, i.e., the HIMEN system.[23,24]

In Memoriam

I.A. Gontova passed away during the production of this chapter.

References

1. Bianki VI. The characteristic of interhemispheric asymmetry. In: Batuev AS, ed. Mechanism of Paired Brain. Leningrad: Nauka, 1989:43-54.
2. Bizarra A, Rogers IJ, Vallortigara CI. The origins of cerebral asymmetry: A review of evidence of behavioral and brain lateralization in fishes, reptiles and amphibians. Neurosci Biobehav Rev 1998; 22:411-426.
3. Nikolova P, Stoyanov Z, Negrev N. Functional brain asymmetry, handedness and menarcheal age. Int J Psychophysiol 1994; 18:213-215.
4. Graves R. Mouth asymmetry, dichotic ear advantage and tachistoscopic visual field advantage as measures of language lateralization. Neuropsychologia 1983; 21:641-649.
5. Bryden MP. Perceptual asymmetry in vision: Relation to handedness, eyedness, and speech lateralization. Cortex 1973; 9:419-435.
6. Biddle FG, Eales BA. The degree of lateralization of paw usage (handedness) in the mouse is defined by three major phenotypes. Behav Genet 1996; 26:391-406.
7. Vant Ent D, Apkarian P. Inter-hemispheric lateralization of event related potentials: Motoric versus nonmotoric cortical activity. Electroencephalogr Clin Neurophysiol 1998; 107:263-276.
8. Soros P, Knecht S, Imai T et al. Cortical asymmetries of the human somatosensory hand representation in right- and left-handers. Neurosci Lett 1999; 271:89-92.
9. Rosen GD. Cellular, morphometric, ontogenetic and connectional substrates of anatomical asymmetry. Neurosci Biobehav Rev 1996; 20:607-615.
10. Vartanyan GA, Klement'ev BI. The chemical induction of central motor asymmetry. In: Kruglikov RI, ed. Chemical Symmetry and Cerebral Asymmetry. Leningrad: Nauka, 1991:5-24.
11. Grebenshikov AYU, Poveshchenko AF, Abramov VV et al. Expression of IL-1 beta gene in brain after peripheral administration of thymus-dependent and thymus-independent antigens. Dokl Biol Sci 1999; 366:294-296.
12. Gerendai I, Halasz B. Neuroendocrine asymmetry. Front Neuroendocrinol 1997; 18:354-381.
13. Sullivan RM, Gratton A. Lateralized effects of medial prefrontal cortex lesions on neuroenocrine and autonomic stress responses in rats. J Neurosci 1999; 19:2834-2840.
14. Neveu PJ. Cerebral lateralization and the immune system. Int Rev Neurobiol 2002; 52:303-23.
15. Neveu PJ. Brain-immune cross-talk. Stress 2003; 6:3-4.
16. Abramov VV, Abramova TYA. Functional asymmetry of immunocompetent cells from contralateral lymphoid organs. In: Komarova LB, ed. Nervous, Endocrine, and Immune Systems Asymmetry. Novosibirsk: Nauka, 1996:44-57.

17. Abramov VV, Kozlov VA, Karmatskich OL. The asymmetry of exogenous CFUs-12 forming in mice. In: Korneva EA, Polyak AI, Frolov BA, eds. The Interactions between the Nervous and Immune Systems. Leningrad, Rostov-na-Donu: Publishing House im. MI Kalinina, 1990:172.
18. Abramov VV. The functional system - a basis of integration of immune and nervous systems. The dissertation of Doctor Medical Sciences. Complex Mechanisms of Interactions between the Immune and Nervous Systems. Moscow: 1991:116-131.
19. Abramov VV, Konenkov VI, Gontova IA et al. Asymmetry of phenotypical and functional characteristics of the cells from lymphoid organs. Dokl Rossiisk Akad Nauk, 1992; 322:802-805.
20. Gontova IA, Abramov VV, Kozlov VA. Asymmetry in cerebral hemispheres and thymus lobes during realization of humoral immune response in mice. Byull Eksp Biol Med 2001; 131:64-66.
21. Gontova IA, Abramov VV, Kozlov VA. Lateralization of thymus lobes and immune response in (CBA × C57Bl/6) F_1 mice. Immunologiya 2000:2:30-32.
22. Abramov VV, Gontova IA, Kozlov VA. Functional asymmetry of the brain and bone marrow in hemopoiesis in (CBA × C57Bl/6) F_1 mice. Byull Eksp Biol Med 2002; 133:468-470.
23. Abramov VV, Gontova IA, Kozlov VA. Functional asymmetry of thymus and the immune response in mice. Neuroimmunomodulation 2001; 9:218-224.
24. Gontova IA, Abramov VV, Kozlov VA. The role of asymmetry of nervous and immune systems in the formation of cellular immunity of (CBA × C57Bl/6) F_1 mice. Neuroimmunomodulation 2004; 11:385-391.
25. Bures J, Buresova O, Huston JP. Techniques and Basic Experiments for the Study of the Brain and Behavior. 2rd ed. Amsterdam: 1983.
26. Till JE, McCulloch EA. A direct measurement of the radiation sensitivity of normal mouse bone marrow cells. Radiat Res 1961; 14:213-222.
27. Yoshikai Y, Miake S, Matsumoto T et al. Effect of stimulation and blockade of mononuclear phagocyte system on the delayed footpad reaction to SRBC in mice. Immunology 1979; 38:577-583.
28. Bulloch K. The comparative study of autonomic nervous system innervation of the thymus in the mouse and chicken. Int J Neurosci 1988; 40:129-140.
29. Bulloch K, McFwen BS, Nordherg J et al. Selective regulation of T-cell development and function by calcitonin gene-related peptide in thymus and spleen. An example of differential regional regulation of immunity by the neuroendocrine system. Ann NY Acad Sci 1998; 840:551-562.
30. Kennedy B, Ziegler MG, Shannahott-Khalsa DS. Alternating lateralization of plasma catecholamines and nasal patency in human. Life Science 1986; 38:1203-1214.
31. Hachinski VC, Oppenheimer SM, Wilson JX et al. Asymmetry of sympathetic consequences of experimental stroke. Arch Neurol 1992; 49:647-702.
32. Wittling W, Block A, Genzel S et al. Hemisphere asymmetry in parasympathetic control of the heart. Neuropsychologia 1998; 16:461-468.
33. Tarkowski E, Naver H, Wallin BG et al. Lateralization of T-lymphocyte responses in patients with stroke. Effect of sympathetic dysfunction? Stroke 1995; 26:57-62.
34. Tarkowski E, Jensen C, Ekholm S et al. Localization of the brain lesion affects the lateralization of T-lymphocyte dependent cutaneous inflammation. Evidense for an immunoregulatory role of the right frontal cortex-putamen region. Scand J Immunol 1998; 17:30-36.

CHAPTER 13

Relation of Behavioral Asymmetry to the Functions of Hypothalamus-Pituitary-Adrenal and Reproductive Systems in Vertebrates

Larissa Yu Rizhova,* Elena Vershinina, Yurii G. Balashov, Dmitri A. Kulagin and Elvina P. Kokorina

Abstract

It is well known that the cerebral hemispheres are involved in the regulation of motor systems and modulation of perceptual cues coming from the contralateral side of the body, and unilateral motor and sensory activity can feedback to brain asymmetry. Recent data also suggest that the right and left cerebral hemispheres differ in their ability to regulate autonomic processes, and direct unilateral stimulation of the brain provokes side-dependent endocrine, immune and other visceral reactions. An important theoretical and practical question of whether or not autonomic processes can be asymmetrically regulated via the activation of the lateral behavioral reactions remains unexplored. In this study, we report that the chronic presentation of an important stimulus—food—from the left side, improves reproductive performance in both cows having normal and poor feed. The unilateral presentation of food can also influence lactation, but in this case the side-dependent effects are different under varying feeding conditions. In the laboratory experiments on rats asymmetrical feeding provoked side dependent changes in dynamic of corticosteroids and gastric ulcer development. Given the underlying mechanisms are similar in all vertebrates including man, these results suggest a simple practical approach of influencing basic somatic functions and have broad applications in agriculture, medicine, ergonomics, and other fields of human activity.

Introduction

Vertebrate brain consists of two laterally symmetrical hemispheres that are involved in regulation of virtually all organism functions. Although both the right and the left hemispheres regulate cognitive abilities, emotions, motivation, muscle activity, and visceral processes, each hemisphere is uniquely dominant for different types of information processing and behaviors. For example, it is widely accepted that the left hemisphere is dominant for formal and analytical thinking, for the sequential ordering of sensory processing, and for the control of speech formation.[1] In addition, the left hemisphere specializes in monitoring time-related characteristics

*Corresponding Author: Larissa Yu Rizhova—Department of Physiology and Biochemistry of Lactation, All-Russia Research Institute for Farm Animal Genetics and Breeding, Moskovskoe Shosse 55-A, 196625, St. Petersburg-Pushkin, Russia. Email: breusch@pc.dk

Behavioral and Morphological Asymmetries in Vertebrates, edited by Yegor B. Malashichev and A. Wallace Deckel. ©2006 Landes Bioscience.

of the events. Contrary, the right hemisphere is dominant in its processing of spatial character-istics,[2] and provides holistic, visual and metaphorical thinking, and Gestalt-like perception, and controls emotional aspects of speech (i.e., prosody).[1,3,4] These functional specializations of the cerebral hemispheres are called "brain asymmetry" actively studied for about 150 years. Only since 1960 till the end of the XXth century about 7000 papers have been published in this field of enquiry.[5]

Compared to brain asymmetry research in the control of psychic processes and behavior, the research of brain hemispheres specialization in regulation of visceral functions is lagging considerably behind. However, for the last 20 years a series of publications have indicated that there is asymmetrical control by the brain of the autonomic nervous system,[6-8] cardiovascular system,[9-11] immune system,[12,13] and endocrine organs, including gonads, adrenals, and the thyroid gland.[14-17] Furthermore, peripheral endocrine gland feedback asymmetrically to the cerebral hemispheres via a neural[18-20] or an endocrine[21,22] route. Numerous observations dem-onstrating brain asymmetry in the regulation of visceral functions exist.[23] For example, the right hemisphere dominates the regulation of heart rate, whereas the left hemisphere is more effective in controlling myocardial contractility. The left hemisphere seems to enhance the responsiveness of several T-cell dependent immune parameters. The right hemisphere, in this case, activates immunosuppressive reactions.

Hormonal and stress-related responses are also controlled by the asymmetrical brain. For example, animals subjected to lesion of the right or left medial prefrontal cortex exhibit different performance of foot shock escape and behavioral asymmetry,[24] and respond differ-ently to stress in their level of plasma corticosterone, gastric ulcer pathology and defecation during a stressful procedure.[25] Additionally, hormonal changes that occur during the estrus circle or experimentally provoked endocrine disorders are accompanied by changes in behav-ioral asymmetry.[26-28]

Humans and nonhuman animals with strongly lateralized hand or side preferences (i.e., right or left handedness or body turns clockwise/anticlockwise) exhibit numerous differ-ences in the functioning of autonomic processes such as susceptibility to different somatic diseases[29,30] or variations in the hormonal reaction to stress.[31,32] The behavioral asymmetry can be activated by training, which, in turn, causes lateral morphological and biochemical changes in the brain.[33-35] These observations suggest that the left and right cerebral hemi-spheres, and the associated behavioral and visceral functions they control, are organized into functional systems.

Based on the above data, we hypothesize that activation of lateral behavioral reactions can provoke corresponding side-dependent visceral changes. To date, data supporting this link between the behavioral asymmetry and visceral responding are missing. Here we show that chronic activation of behavioral asymmetry in animals, as produced by lateralized presenta-tion of meaningful stimuli, provokes changes in reproductive abilities, lactation, dynamics of corticosteroids during a stress procedure, and erosions on the stomach mucous.

Materials and Methods

The Experiments on Cows

One of the traditional methods used in Russian intensive farming to keep cows produc-ing milk is to arrange the animals in the cattle yards so that they are oriented asymmetrically in relation to their food, the most important stimuli for them. The cows are placed in two opposite rows with a passageway for fodder between the rows. If the fodder always moves in the same direction, then the animals standing in one row see the food coming from the right side (R-cows) while animals in the opposite row see the stimuli first appeared from the left side (L-cows). This manner of feeding is especially suitable for examining the role of asym-metrical (left vs. right-sided) food presentation on somatic responses. Thus, all the subjects studied in this experiment were cows housed on two large-scale farms with herd sizes of approximately 1000. Each cow had an individual registration form recording information about its monthly milk production, milk composition (percent of protein and fat), and the dates of successive inseminations and confinements until the day the subject was culled. The cows were artificially inseminated in the estrus period, as soon as the estrus cycle resumed after calving. If the insemination was not successful, it was repeated at the onset of the next estrus cycle and so on until the cow became pregnant. The duration of the period between calving and the next pregnancy, a so called "service period," is a widely used measuring for evaluation the reproductive abilities in cows. The dates of all inseminations of individual animals, information about the bull whose sperm was used and the results of the inseminations were registered in a special journal in the farm.

The experiments were conducted on Holstein dairy cows milked with the vacuum milking machines connected to the pipeline on the Rizhiki and Happo-Oye industrial farms in Leningrad province of Russia.

Rizhiki Farm

In this farm the animals had a standard daily diet and a stable annual milk production of about 4500–5000 litres per cow for about ten years. The same workers milked and fed all the animals in one row, and their names was recorded on the cow registration forms. The cow forms of the regular farmhands that had worked on the same row for at least eight years were found in the farm archive in order to determine where each cow stood. We used information about the animals covering the period from the second lactation until the culling date, because the cows were allocated to fixed positions after their second calving on this farm. The workers delivered food to the animals from one side for only four months between February and May each year, so there were two rows containing L- and R-cows in each cattle yard during this period. In summer the animals were taken out to pasture, and they came to their own places in the cattle yards three times a day for milking and some additional feeding. From June to January, the animals were fed from both sides. The animals standing in both rows were given equal amount of food.

Happo Oye Farm

The Happo Oye farm experienced economic problems during the year of the experiment. The amount of concentrated mix in the animals' daily diet was decreased by 50% in comparison with a stable period, and the level of annual milk production decreased to 3000 liters per cow. The feeders always moved in one direction. The same worker milked and fed 25 animals in one row and 25 animals in the opposite row. We selected L- and R-cows, which calved in October–November and observed them for seven months.

The Experiments on Rats

Methods

Animals

One month old female Wistar rats were taken from their mothers and housed in groups of six per cage (40 × 60 × 20 cm). They were maintained under natural light/dark conditions at a constant temperature (22 ± 1°C) and fed a standard diet. The cages had an additional wall separating a narrow compartment 40 × 7 cm. Food and water were placed at one end of the compartment. The rats could enter the compartment through a hole (0.5 cm) in the middle of the wall. The subjects were divided into three groups, which differed in the direction they had to turn in order to reach food and water. These included:

- rats, which always had to turn right (R-rats);
- rats, which always had to turn left (L-rats);
- rats, which alternated turning left one day, and right next day (ambilateral, A-rats).

To motivate the animals to go for food the feeder was emptied at 8-00 pm, and filed up again at 8-00 am. The animals lived under these conditions for two months before the testing began, and till the end of the experiments.

Test for Behavioral Asymmetry

The animals were given two tests: "handling by the tail" and T-maze. The "handling by the tail" test (TH-test) was modified from Castellano et al.[36] A rat was removed from her cage, grasped by her tail about 4 cm from the end, and lifted up until it was about 30-40 cm above a tabletop. The animal's left/right direction of lateral body flexion or rotation was noted as she tried to right herself. Immediately after this, the animal was placed into a T-maze. The stem of the maze was a pipe with a diameter of 8 cm. The cross-piece of the maze was 100 × 12 cm. When the rat was released at the stem of the maze, a stopwatch was started. When the animal started moving towards the end of the alley the stopwatch was stopped and the elapsed time was recorded (latency). The rat was given a maximum of 3 min for this procedure. If the animal did not start to move into the alley within this time it was forced to do so by giving it a light pinch on its tail. As soon as the rat came to the end of the alley it was handled and returned to the home cage.

The procedure described was performed ones a day for eight successive days. For each animal the proportion of left turns was calculated and used as an index of laterality bias.

Stress Procedure

After eight days of behavioral testing the animals were subjected to a stress procedure. The strong stress was necessary to induce ulcer formation on the stomach mucous, and to measure stress resistance in the animals. A strong immobilization stress was employed in Pavlov Institute of Physiology in St. Petersburg 15 years ago when these experiments were performed. The animals were tied to a desk by all four paws flat on their bellies. The temperature in the room was 10° C. Blood samples were taken from the animal's tail after two-three minutes, 30 min, and one, two and four hours, to determine the concentration of corticosteroids. The duration of the stress was 24 hours. The animals were then decapitated, and a final blood sample was taken. Their stomachs were dissected, washed carefully and smoothed out on a table to measure the area of erosions. For this purpose, a transparent net with square cells (1 × 1 mm) was placed on the stomach mucous. The measurement was performed visually.

Determination of the Blood Corticosteroids by Sulphuric Acid-Ethanol Fluorescence Derivatization[37]

0.1 ml of blood plasma was mixed with 1.5 ml of methylene chloride by shaking for 3 min, and frozen in dry ice-ethanol bath for 2 min. Afterwards the methylene chloride extract was transferred into a new tube and mixed with 0.5 ml of acid-ethanol mixture consisting of seven parts of sulphuric acid and three parts of absolute ethanol. After mixing by tapping the tube for three minutes, the upper layer of methylene chloride was gently removed by suction. After 1.5 hours a Hitachi F4500 spectrofluorimeter was used to measure, at ambient temperature, the fluorescence emission from a 0.5 cm quartz cuvette at an emission wavelength of 520 nm (excitation wavelength of 365 nm). The concentration of corticosteroids in plasma samples was determined from a standard curve.

Data Analysis

Student's *t*-test and Mann-Whitney U-test were used to compare characteristics of two groups of the animals with left- and right-side orientation to food. Chi-squared statistics was used to compare the data in percentage, one-way ANOVA and Scheffe post hoc tests were employed to verify a hypothesis that the factor "asymmetry in orientation to food" having three gradations (sinistral, dextral and ambidextral) influences behavioral and hormonal characteristics in the animals.

Results

The Experiments on Cows

Presentation of the fodder always provoked a reaction in the animals, even if they still had food. Specifically, cows turned their heads toward the food and exhibited increased general activity.

The Influence of Lateral Presentation of Food on Reproduction and Lactation in Cows Living under Normal Feeding Conditions (Rizhiki)

To determine whether lateralized presentation of food influenced reproductive performance in the cows, we compared the duration of the "service period" for animals that stood in opposite rows, or in the same rows, after calving in different months of the year as a function of feeding asymmetry. For this purpose, we located all the cases when the animals calved in, for example, January, in the registration forms for L- and R-cows, and extracted the duration of "service period" after these calves. The same procedure was followed for the calves in all other calendar months. Since the individual animals delivered calves in different months in different years, the data were segregated in the corresponding groups.

The cows had between two and eight calves. We were, unfortunately, unable to compare the data on the same cows appeared in different years during periods of asymmetrical versus symmetrical feeding, as the duration of "service period" depends on a number of factors including the age of the cows, quality of food, season, quality of the sperm (the bull whose sperm was used) and level of lactation. The individual cows standing in the same row had a different set of these factors at the moment of their successive inseminations in different years. The asymmetrical and symmetrical presentation of food took place in different seasons of the year. Conversely, the cows standing in opposite rows always had a few identical factors (sperm of the same bull, feed, season, and age distribution). The comparison of "service periods" for cows standing in opposite rows in different months of the year using Student's *t*-test and Mann-Whitney U-test was therefore a valid measure.

As can be seen in Figure 1, after calving, during the period when the feeders brought food from only one side, L-cows had significantly shorter service periods than R-cows. This difference disappeared in the cows in the same rows after calving between June and December, when food was delivered from both sides. As it is seen in the Figure 1 the cows standing in the opposite rows and calving in January, i.e., in the last month of symmetrical feeding, had significantly different "service period" like L- and R-cows. It happened because cows usually begin to cycle about 40-50 days after calving and therefore the animals calving in January could not be inseminated earlier than February when feeding was asymmetrical. Conversely, L- and R-cows calving in May, i.e., in the last month of asymmetrical feeding, began cycling in June when they grazed, and some additional food was delivered from both sides. Nevertheless they kept the difference in duration of "service period" provoked by asymmetrical feeding. An argument can be adduced to explain this contradiction in the results. In May the animals exhibited improvement of reproductive performance depending on the season, and the duration of "service period" decreased in all the animals compared to winter months. The difference was significant in R-cows (108.80 ± 8.40 days in January compared to 87.96 ± 6.19 days in May, $p < 0.05$). Perhaps, the effect of asymmetrical feeding on the reproductive abilities in the animals depends on the level on which the reproductive system is functioning: the more intensive is the functioning the stronger the effect of the external asymmetry is, and the shorter the exposition to asymmetrical feeding can be to induce significant changes.

It is well known that the first two months after calving are very important for the development of milk gland function, since it is during this period that cows increase their lactation to the maximum level. We compared the lactation curves for L- and R-cows, which calved in April and hence received food from only one side for about two months at the beginning of their lactation. As can be seen in Figure 2A, the L-cows produced more milk

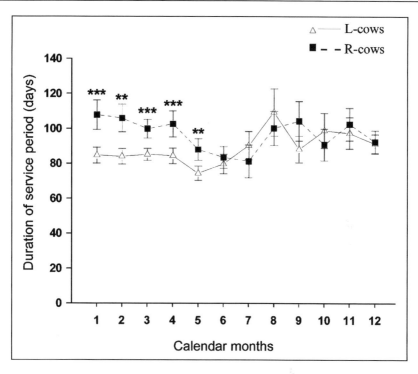

Figure 1. Duration of the service period for L- and R-cows after calving in different months of the year, Rizhiki. Between February and May, fodder appeared from the left side for L-cows (Δ), and from the right side for R-cows (■). Between June and January, fodder came to the animals from both sides. Data are given as means ± SEM. The number of animals was 1,135 (45-175 in different months) and 811 (39-122 in different months) for L and R-rows respectively. **: p < 0.01, ***: p < 0.001 when compared with L-cows, Mann-Whitney *U*-test. (Reproduced from Rizhova et al[55] with permission from Elsevier Science).

than the corresponding R-cows, and they maintained a higher level of production even after food began to arrive from both sides. Cows that stood in the same rows after calving in December (and received food from both sides for the two first months of lactation or less) did not differ in their milk production even once the period of asymmetry had begun (Fig. 2B).

The milk gland in cows develops with the number of lactations. Table 1 shows milk production for cows of different ages, placed in opposite rows. A significant difference was seen only in the period when food was delivered from one side in cows with three or more lactations. The L- and R- cows with two lactations, producing less milk than the older animals, did not differ significantly. This shows that the effect of asymmetrical feeding on lactation is stronger when the milk gland is functioning on a higher level. Animals of all ages standing on the same places in the opposing rows did not differ during the period when food was delivered from both sides. This observation is likely due to the effect of lateralized food presentation rather than other factors, such as row placement, the presence of other animals, etc., all of which were randomly distributed among subjects.

The Influence of Lateral Presentation of Food on Reproduction and Lactation in Cows Living under Conditions of Poor Feed (Happo Oye)

As we have mentioned above, the farm Happo Oye experienced economical difficulties and the animals had poor feed during the year of the experiment. As we have shown earlier in the experiments on heifers, the animals changed behavioral asymmetry when the living conditions

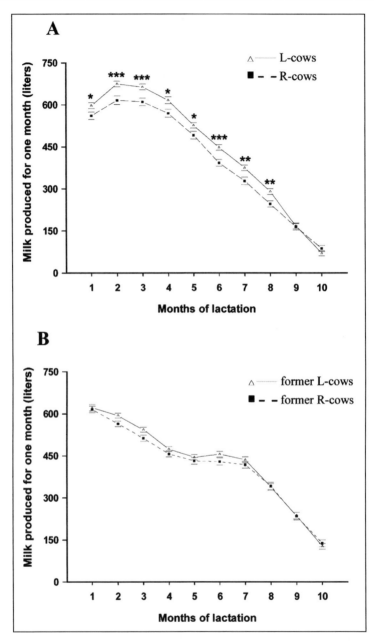

Figure 2. Lactation curves for cows with asymmetrical and symmetrical orientation to food at the beginning of lactation, Rizhiki. A) Monthly milk production for cows which calved in April when the feeders brought food to the animals from one side only, during the first two months of lactation. The fodder appeared from the left side for L-cows, N = 135, and from the right side for R-cows, N = 92. Data are given as means ± SEM. *: $p < 0.05$, **: $p < 0.01$, ***: $p < 0.001$ when compared with R-cows, Student's *t*-test. B) Monthly milk production for the cows, which calved in December when the fodder appeared from both sides during the first two months of lactation. 224 former L-cows and 145 former R-cows. (Reproduced from Rizhova et al[55] with permission from Elsevier Science.)

Table 1. *Milk production for the whole lactation in the cows of different ages after calving in the period of asymmetrical and symmetrical orientation to food*

Age of Cows (Number of Lactations)	L-Rows		R-Rows	
	N	Milk Production (Liters)	N	Milk Production (Liters)
	Asymmetrical orientation to food			
2	28	4,144.86 ± 99.49	17	3,940.29 ± 176.19
3-4	61	4,534.02 ± 91.03	43	4,115.51 ± 124.32**
5-6	25	4,436.24 ± 150.62	24	4,013.71 ± 197.60*
	Symmetrical orientation to food			
2	47	4,165.32 ± 160.47	38	4,231.29 ± 132.74
3-4	86	4,420.94 ± 91.03	58	4,401.51 ± 112.72
5-6	50	4,220.70 ± 133.73	28	4,118.18 ± 188.25

In the period of asymmetrical orientation to food the fodder brought food to the animals from one side only (from the left side to the cows in L-row, and from the right side to the cows in R-rows). In the period of symmetrical orientation to food the fodder appeared from both sides. Data are given as means ± SEM. *: $p < 0.05$, **: $p < 0.01$ when compared to the animals in L-rows. (Reproduced from Rizhova et al[55] with permission from Elsevier Science.)

changed for the worse.[38] In this case the unilateral activation of the animals by delivering food from one side could provoke different effects on the reproduction and lactation.

To compare "service period" in L- and R-cows we grouped them based on the number of calves they had had, since the "service period" could decrease for cows as they had more calves. As can be seen in Figure 3, in all groups, L-cows had significantly shorter "service period" than corresponding R-cows. Thus, the impairment of feeding conditions did not change the effect of asymmetrical feeding on reproductive performance that was seen in the animals having normal feed.

Comparing milk production in L- and R-cows under poor feeding conditions, we found that L-cows were not superior in this case. The amount of milk produced for seven months was 2830.19 ± 93.22 and 3036.76 ± 84.00 litres, in L- and R-cows respectively. The difference was not significant, though R-cows had a nonsignificant trend for continued lactation compared to L-cows (see Fig. 4). The results thus indicate that either left or right side orientation to food can be beneficial for lactation in the animals, under varying feeding conditions.

The Experiments on Rats

It has been shown that individuals with different behavioral asymmetry differ in their ability to adapt.[39,40] Can the adaptation processes be changed using a kind of behavioral asymmetry training, and will this influence change other lateral biases in the individuals? This has not been investigated. To examine this, we exposed rats to a period of asymmetrical feeding and then examined the effect of asymmetrical feeding on measures of behavioral asymmetry, anxiety and on the response to acute stress.

When grasped by the tail and lifted up, rats perform lateral body flexions or rotations. The pattern of this behavior has been described in detail in our earlier publication.[27] One-way ANOVA revealed that the direction the animals had to turn to reach food had a significant effect on their lateral bias in the TH-test (F $(2,70) = 6.355$, $p = 0.003$). Post hoc testing revealed that the probability of left-side body movements was higher in L-rats than in R-rats ($p = 0.012$). A trend was seen for A-rats to make less left-side movements than L-rats ($p = 0.06$; see Fig. 5). No difference was found between the groups in their lateral bias in T-maze. The probability of a left turn was 0.48 ± 0.05 in L-rats, 0.41 ± 0.04 in R-rats, and 0.45 ± 0.03 in A-rats. The

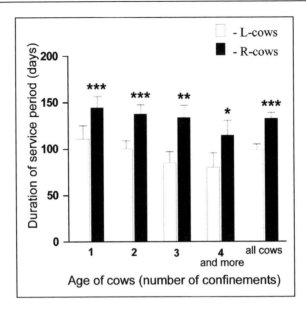

Figure 3. Duration of the service period for cows of different ages with asymmetrical orientation to food, Happo-Oye. The fodder appeared from the left side for L-cows, N = 90 (10-40 in different groups), and from the right side for R-cows, N = 99 (11-35 in different groups). Data are given as means ± SEM. *: p < 0.05, **: p < 0.01, ***: p < 0.001 when compared to L-cows, Mann-Whitney U-test. (Reproduced from Rizhova et al[55] with permission from Elsevier Science.)

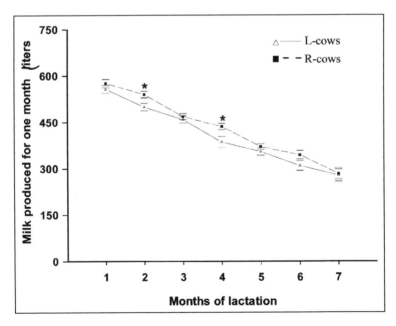

Figure 4. Lactation curves for cows with asymmetrical orientation to food living under poor feeding conditions, Happo-Oye. The fodder appeared from the left side for L-cows, N = 90, and from the right side for R-cows, N = 99. Data are given as means ± SEM. *: p < 0.05 when compared to L-cows, Student's t-test. (Reproduced from Rizhova et al[55] with permission from Elsevier Science.)

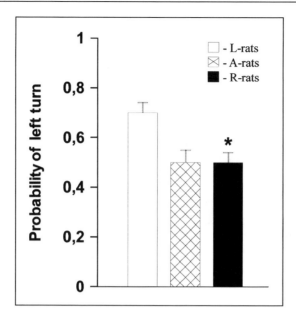

Figure 5. Lateral bias in TH-test in the rats with asymmetrical orientation to food. L-rats had to turn to the left to get food in the home cage, N = 25, R-rats had to turn to the right, N = 25, and A-rats had to turn one day to the right and next day to the left, N = 23. Data are given as means ± SEM. *: p < 0.05 when compared to L-rats, ANOVA.

latency in T-maze was measured for individuals as an average mean value for eight trials. The distribution of this characteristic in the population was different in L-, R-, and A-rats (Fig. 6). ANOVA revealed significant effect of the factor "asymmetry in orientation to food" on the latency in T-maze (F (2,70) = 5.224, p = 0.008). Post hoc testing revealed that the latency was longer for L-rats than for R-rats (p = 0.012; Fig. 7), while a trend was seen for A-rats to have shorter latency than L-rats (p = 0.06). Since similar latency values were observed for all animals in the first trial (15.87 ± 3.21 sec, 16.76 ± 5.87 sec and 18.35 ± 4.21 sec in L-, R- and A-rats respectively) the increase in latency for the L-rats in the subsequent trials could be a result of higher anxiety, provoked by handling at the end of the T-maze alley. To investigate this, we returned to the data of our previous experiments[27,28] on the female Wistar rats that were tested in the same experimental design (T-maze after TH-test), but which were not handled in T-maze. In that experiment both alleys had exits to the dark transport cages, which the rat could enter. We compared the average latency value for eight trials in the T-maze for rats, which made 0-1, 3-5, or 7-8 left turns in the TH-test. No significant difference was found, although the rats, which made 7-8 left turns in TH-test tended to have a shorter latency (19.32 ± 3.00 sec, N = 87) compared to rats, which made 0-1 (29.35 ± 7.45 sec, N = 44) and 3-5 left turns (20.33 ± 5.94 sec, N = 37). Taken together, the results of both experiments show that L-rats had a greater response to handling in T-maze and anxiety compared to R-rats, and that this caused them to stay longer in the stem of the maze.

The dynamics of corticosteroids in blood during immobilization stress is seen in Figure 8. During the first 30 min of the stress, the rats of all three groups had similar levels of corticosteroids, after which time a decrease was observed in R- and A-rats. ANOVA revealed a significant effect of the factor "asymmetry in orientation to food" on the level of corticosteroids for four hours after the stress began (F(2,30) = 3.455, p = 0.045). Post hoc testing showed that R-rats had lower hormonal level at this time point (p = 0.05), compared to L-rats. Student's t-test, employed to do group comparisons, demonstrated a significant difference in corticosteroid

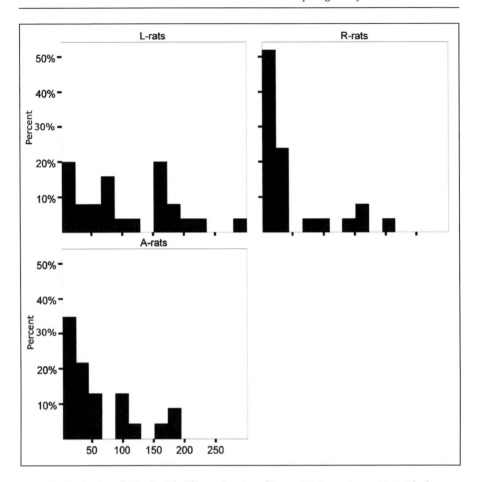

Figure 6. Distribution of animals with different duration of latency in T-maze. L-rats, N = 25, had to turn to the left to get food in the home cage; R-rats, N = 25 had to turn to the right, and A-rats, N = 23 had to alternate turning to the left one day, and the right next day.

levels between L- and R-rats beginning one hour after the onset of stress ($p < 0.05$). The data thus showed that L- and R-rats had different dynamics of corticosteroids during the stress. L-rats had a stronger response of hypothalamic-pituitary-adrenocortical axis and constantly high level of corticosteroids during at least four first hours of the stress. Conversely, the hormonal reaction to stress in R-rats decreased with time. These animals could respond to the additional stress provoked by handling and decapitation after 24 hours: the level of corticosteroids in blood in R-rats increased and was significantly higher than in L-rats ($F(2,48) = 3.534$, $p = 0.037$, ANOVA and $p = 0.05$, post hoc). The results showed that asymmetrical feeding induced essential side-dependent changes in the function of hypothalamic-pituitary-adrenocortical axis.

ANOVA did not reveal any significant effect of "asymmetry in orientation to food" on the area of erosions on the stomach mucous. Student's two-tailed t-test revealed that the area of erosions was larger for L-rats than for R-rats ($t = 2.194$, $p = 0.042$; Fig. 9), indicating greater stress-resistance in R-rats under conditions of an acute immobilization stress than L-rats.

The results of these experiments on rats thus indicate that manipulations, which force the rats to feed asymmetrically, cause side-dependent changes in their behavioral and visceral functions. Specifically, these manipulations cause changes, on lead to behavioral asymmetry in

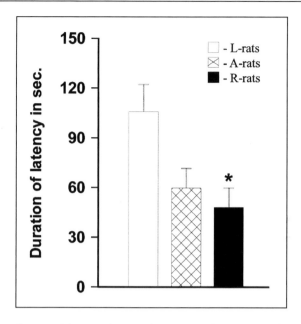

Figure 7. Duration of latency in T-maze in the rats with asymmetrical orientation to food. L-rats had to turn to the left to get food in the home cage, N = 25, R-rats had to turn to the right, N = 25, and A-rats had to turn one day to the right and next day to the left, N = 23. Data are given as means ± SEM. *: p < 0.05 when compared to L-rats, ANOVA.

the TH-test, result in different performance of anxiety related behavior, cause changes in the way how hypothalamic-pituitary-adrenocortical axis responded to stress, and degree of erosions on the stomach mucous as a measure of stress resistance.

Discussion

The experiments on the cows have shown that activation of behavioral asymmetry in animals by presenting food from different sides influenced their reproduction and lactation. To our knowledge, this is the first study demonstrating the possibility of stimulating reproduction and lactation via the behavioral asymmetry route. The last review of the reproductive challenges facing the cattle industry at the beginning of this decade does not mention the possible involvement of brain and behavioral asymmetry.[41]

The cows standing in the same, but opposing, rows in the cattle yard differed in their lactation and reproduction characteristics only when they saw the food coming from the left or from the right. The same animals showed no differences when the food was delivered from both sides. The experiments on rats indicate that the animals turning to the left or to the right to get food in their home cages differed in their hormonal reaction to stress, and the degree of erosion of the mucous of stomach. These facts prove that the lateral position of meaningful stimuli provoking lateral sensory reactions and body turns influences somatic functions in animals.

A number of published reports show side-biases in the control of feeding responses in vertebrates.[42] Studies on chicks have shown that using the right eye helps them to discriminate between grains and pebbles.[43] Toads demonstrate feeding responses to prey more readily when it is located in the right visual hemifield.[44,45] Nevertheless, this right-side preference in the feeding responses is not in contradiction to our results. The dominance of the right eye and left hemisphere in the control of feeding responses is not absolute. The right eye is preferred in cases where a decision has to be made whether to bite or not, or to inhibit a reaction to an

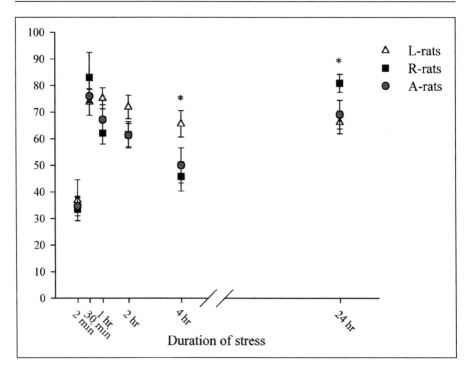

Figure 8. Dynamic of corticosteroids in blood in the rats with asymmetrical orientation to food during acute immobilization stress. L-rats had to turn to the left to get food in the home cage, N = 11 to 16 at different time points, R-rats had to turn to the right, N = 12 to 17, and A-rats had to turn one day to the right and next day to the left, N = 10 to 15. Data are given as means ± SEM. *: p < 0.05 when L- and R-rats are compared, ANOVA.

inedible target,[46] and in cases where an object has to be manipulated to get to food, for example, removing a lid from the dish containing food.[47] When discrimination or manipulation is not necessary, for example, when the food dish is not covered with a lid, the left eye is preferred.[46] The cows in the cattle yard and the rats in the cages do not have to distinguish their food from inedible objects. We have shown earlier that two to four-month-old heifers have a populational bias to turn the head left to get food. When presented with two identical plates of food placed symmetrically in front of the head, the majority of the animals preferred to eat from the left plate first.[48] We think that the experimental models employed for the studies of perceptual asymmetry in vertebrates mentioned above differ principally from the models we used in our experiments. In the former, the animals might be ready to react to food appearing with the same probability in the left or in the right visual hemifield, or be able to react to food in a mix of edible and inedible objects. Having an asymmetrical brain, they performed postural reactions and movements to facilitate the most efficient perception and direct transmission of information to the hemisphere specialized in responses to this kind of tasks. The cows and rats in our experiments saw food constantly appearing or placed on one side. They could not change the position of their bodies in relation to food as they perhaps may have wanted to, since they were tied to their places or moved in a narrow passage. In these situations, one side of the space around the animal became more meaningful, and brain asymmetry was brought in correspondence with the asymmetry of the space. From this point of view our model is closer to the models where direction of functional asymmetries was changed by monocular deprivation[49,50] or by training the unilateral motor activity.[33,51]

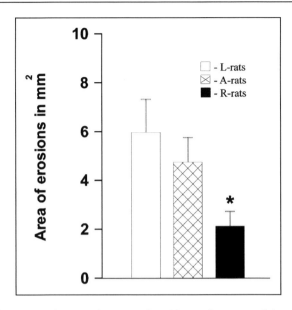

Figure 9. Area of erosions on the stomach mucous after 24 hours of acute immobilization stress in the rats with asymmetrical orientation to food. L-rats had to turn to the left to get food in the home cage, N = 14, R-rats had to turn to the right, N = 12, and A-rats had to turn one day right and next day left, N = 15. Data are given as means ± SEM. *: p < 0.05 when compared to L-rats, Student's *t*-test.

A series of publications have indicated that the stimulation of behavioral and perceptual asymmetry activates the contralateral brain hemisphere. For example, stimulating the animals to perform unilateral body turns in a conditioned circling paradigm resulted in an increase of 3,4-dihydroxyphenylacetic acid/dopamine ratio, reflecting greater electrical activity in the contralateral cortex.[52] Unilateral sensory and manual activity promoted normalization of physiological processes in the contralateral hemisphere after lesion.[53,54] Thus, the food appearing constantly from the left side might activate the right brain hemisphere and create a steady unilateral dominance. Since L- and R-cows were given the same amount of food, and as we have shown earlier,[55] L-cows had longer life and produced more calves than R-cows, maintaining right-side cerebral dominance, therefore, seems to increase the effective utilization of food for lactation and reproduction, rather than leading to self destruction and impairment of health. We have also shown[55] that the time taken for the reproductive cycle to resume after calving, and the duration of the estrus cycle, were similar for L- and R-cows, and so the asymmetry in food presentation did not influence these characteristics. In our laboratory experiments female rats with strong left- and right side lateral bias in T-maze did not differ in the duration of estrus cycle,[28] but proportion of the proestrus stage, was significantly higher in the animals preferring turns to the left. Proestrus stage is critically important for reproduction since this stage differs from other estrus stages in having significantly higher levels of luteinizing hormone, follicle stimulating hormone and estradiol. It is the main period for development of the ovarian follicle.[56] Delivering food to the cows from left side increased the percentage of effective inseminations thus favoring reproduction due to increase the chance of becoming pregnant.[55] Our results, viewed as a whole, indicate that the right cerebral hemisphere seems to dominate regulation of reproduction in cows in a broad range of feeding conditions, and may be generally superior in the control of impregnation. This hypothesis is supported by other findings demonstrating predominance of right side cerebral structures in the control of gonadal function in females.[57,58]

As for lactation, we did not find clear indications of the involvement of brain asymmetry in the regulation of this function. Our data have shown that either the right or left hemisphere could dominate in the control of lactation, depending on the feeding conditions. Activation of the right hemisphere increased the level of lactation in the animals under good feeding conditions, whereas the left hemisphere was more effective when the animals had poor feed. Similar activation of the right brain hemisphere in the rats increased their reaction to handling and anxiety related behavior—latency in the T-maze. These animals had constantly high levels of corticosteroids in their blood during the stress, whereas one hour after beginning the stress the levels of these hormones decreased in the animals which experienced activation of left brain hemisphere. These data corroborate the numerous publications about possible differences between two cerebral hemispheres in neural control of affective reactions and stress response. For example, the rats with left side bias in amphetamine induced circling had a higher level of ACTH in blood when being exposed to a novelty stress than the animals with right side bias.[31] Presenting a three-minute aversive emotional film directly to the left visual hemifield (right brain hemisphere) in human subjects resulted in higher cortisol responses than presenting it to the right visual hemifield (left brain hemisphere).[59] Right hemisphere dominates the control of negative emotions.[60-62] According to the James Henry hypothesis each cerebral hemisphere specializes in monitoring of a specific type of response to stress: active responding and passive responding. Particularly activation of right hemisphere is preferentially associated with the passive behavioral reaction to stress and stronger response of hypothalamic-pituitary-adrenocortical axis.[63] In our experiments the animals subjected to activation of right hemisphere were less stress resistant, since they had larger area of erosions on the stomach mucous than the animals subjected to activation of left hemisphere. Similar results were reported by Sullivan and Gratton.[25] They have shown that lesion of right medial prefrontal cortex decreases stress ulcer development.

Summarizing our results reported in this paper we can conclude that activation of right cerebral hemisphere gives animals a benefit in reproductive performance whereas activation of left hemisphere improves their abilities to cope acute stress. In our mind this specialization of the hemispheres reflects two survival strategies of different species: (1) high level of reproduction and producing great amount of individuals with considerably weak adaptation abilities; (2) low level of reproduction and producing a small number of individuals with high degree of survival. Since these survival strategies are based on functioning of different hormonal systems, we suppose, they might be associated with population asymmetries in different species.

As we have shown the chronic lateral presentation of emotionally meaningful stimuli can influence autonomic processes in the organism, which are regulated by two cerebral hemispheres in asymmetric manner. The mechanisms of brain and behavioral asymmetry are remarkably similar in different species including humans,[64,65] thus the lateral presentation of emotion-related stimuli might affect autonomic processes in other farm mammals, poultry and man. Our research suggests a simple practical approach to influencing basic somatic functions in the organism, and has broad applications for agriculture, medicine, ergonomics and other fields.

References

1. Bradshaw JL, Nettleton NC. The nature of hemispheric specialization in man. Behav Brain Sci 1981; 4:51-91.
2. Dobrochotova TA, Bragina NN. Functional asymmetry and psychopathology of focal lesions of the brain. Moscow, Russia: Meditsina, 1977.
3. Bever TJ. Cerebral asymmetries in humans are due to the differentiation of two incompatible processes: Holistic and analytical. Ann NY Acad Sci 1975; 263:251-162.
4. Chernigovskaya T. Cerebral lateralization for cognitive and linguistic abilities: Neuropsychological and cultural aspects. In: Wind J, Jonker A, eds. Studies in Language Origins. Amsterdam-Philadelfia: John Benjamins Publishing Company, 1994:3:56-76.
5. Beaumont JG. Future research directions in laterality. Neuropsychol Rev 1997; 7:107-126.
6. Hachinski VC, Oppenheimer SM, Wilson JX et al. Asymmetry of sympathetic consequences of experimental stroke. Arch Neurol 1992; 49:697-702.
7. Wittling W, Block A, Genzel S et al. Hemisphere asymmetry in parasympathetic control of the heart. Neuropsychologia 1998; 36:461-468.

8. Wittling W, Block A, Schweiger E et al. Hemisphere asymmetry in sympathetic control of the human myocardium. Brain Cogn 1998; 38:17-35.
9. Kirchner A, Pauli E, Hilz MJ et al. Sex differences and lateral asymmetry in heart rate modulation in patients with temporal lobe epilepsy. J Neurol Neurosurg Psychiatry 2002; 73:3-75.
10. Shapoval LN, Sagach VF, Pobegailo LS. Chemosensitive ventrolateral medulla in the cat: The fine structure and GABA-induced cardiovascular effects. Journal of the Autonomic Nervous System 1991; 36:159-172.
11. Nikolaeva EI, Oteva EA, Nikolaeva AA et al. Prognosis of myocardial infarction and brain functional asymmetry. Int J Cardiol 1993; 42:245-248.
12. Abramov VV, Gontova IA, Kozlov VA. Functional asymmetry of thymus and the immune response in mice. Neuroimmunomodulation 2001; 9:218-224.
13. Bardos P, Degenne D, Lebranchu Y et al. Neocortical lateralization of NK activity in mice. Scand J Immunol 1981; 13:609-611.
14. Banczerowski P, Csaba Z, Csernus V et al. The effect of callosotomy on testicular steroidogenesis in hemiorchidectomized rats: Pituitary-independent regulatory mechanism. Brain Res Bull 2000; 53:227-232.
15. Banczerowski P, Csernus V, Gerendai I. Unilateral paramedian-sagittal brain cut extending from the level of the anterior comissure to the midlevel of the third ventricle above the amygdala affects gonadal function in male rat: A lateralized effect. Acta Boil Hung 2003; 54:79-87.
16. Banczerowski P, Csaba Z, Csernus V. Lesion of the amygdala on the right and left side suppresses testosterone secretion but only left-sided intervention decreases serum luteinizing hormone level. J Endocrinol Invest 2003; 26:29-434.
17. Gerendai I, Halasz B. Neuroendocrine asymmetry. Front Neuroendocrinol 1997; 18:354-381.
18. Bakalkin GYA, Tsibezov VV, Sjutkin EF et al. Lateralization of LHRH in rat hypothalamus. Brain Res 1984; 296:361-364.
19. Gerendai I, Rotsztejn W, Marchetti B et al. LHRH content changes in the mediobasal hypothalamus after unilateral ovariectomy. In: Pollery A, MacLeod RM, eds. Neuroendocrinology: Biological and Cinical Aspects. New York/London: Acad. Press, 1979:97-102.
20. Jones RE, Desan PH, Lopez KH et al. Asymmetry in diencephalic monoamine metabolism is related to side of ovulation in a reptile. Brain Res 1990; 506:187-191.
21. Rodriguez-Medina MA, Reyes A, Chavarria ME et al. Asymmetric calmodulin distribution in the hypothalamus: Role of sexual differentiation in the rat. Pharmacol Biochem Behav 2002; 72:189-195.
22. Xiao L, Jordan CL. Sex differences, laterality, and hormonal regulation of androgen receptor immunoreactivity in rat hippocampus. Horm Behav 2002; 42:327-36.
23. Wittling W. Brain asymmetry in the control of autonomic-physiologic activity. In: Davidson RJ, Hugdahl K, eds. Brain Asymmetry. Cambridge, Massachusetts: MIT Press, 1995:305-358.
24. Carlson JN, Visker KE, Keller RW et al. Left and right 6-hydrixydopamine lesions of the medial prefrontal cortex differentially alter subcortical dopamine utilization and the behavioral response to stress. Brain Res 1996; 711:1-9.
25. Sullivan RM, Gratton A. Lateralized effects of medial prefrontal cortex lesions on neuroendocrine and autonomic stress responses in rats. J Neurosci 1999; 19:2834-2840.
26. Becker J, Robinson TE, Lorenz KA. Sex differences and estrous cycle variations in amphetamine-induced rotational behavior. Eur J Pharmacol 1982; 80:65-72.
27. Rizhova LYU, Kulagin DA. The effects of corticosteroids on lateral bias in female rats. Behav Brain Res 1994; 60:51-54.
28. Rizhova LYU, Vershinina EA. The dynamics of two different tests of laterality in rats. Laterality 2000; 5:331-350.
29. Geschwind N, Galaburda AM. Cerebral lateralization. Biological mechanism, associations, and pathology: I.A Hypothesis and a program for research. Arch Neurol 1985; 42:428-59.
30. Morfit NS, Weekes NY. Handedness and immune function. Brain Cogn 2001; 46:209-213.
31. La Hoste GJ, Mormede P, Rivet JM et al. Differential sensitization to amphetamine and stress responsibility as a function of inherent laterality. Brain Res 1988; 453:381-384.
32. Martins JM, Alves J, Trinca A et al. Personality, brain asymmetry and neuroendocrine reactivity in two immune-mediated disorders: A preliminary report. Brain Behav Immunity 2002; 16:383-397.
33. Greenough WT, Larson JR, Withers GS. Effects of unilateral and bilateral training in a reaching task on dendritic branching of neurons in the rat motor-sensory forelimb cortex. Behav Neur Biol 1985; 44:301-314.
34. Schwarting R, Nagel JA, Huston J. Asymmetries of brain dopamine metabolism related to conditioned paw usage in rat. Brain Res 1987; 417:75-84.
35. Yamamoto BK, Freed AC. Asymmetric dopamine and serotonin metabolism in nigrostriatal and limbic structures of the trained circling rat. Brain Res 1984; 297:115-119.
36. Castellano MA, Diaz-Palarea MD, Barroso J et al. Behavioral Lateralization in rats and dopaminergic system: Individual and population laterality. Behav Neurosci 1989; 103:46-53.

37. Balashov YuG. Fluorimetric micro method for determination of corticosteroids: Comparison with other methods. Physiological Journal of USSR 1990; 76:280-282, (in Russian).
38. Rizhova LYU, Kokorina EP, Filippova EB et al. The influence of living conditions on behavioral asymmetry in heifers. Vestnik Leningradskogo Universiteta 1994; 4:87-96, (in Russian).
39. Collins RL. When left-handed mice live in right-handed worlds. Science 1975; 187:181-184.
40. Iliuchonok RY, Filkenberg AL, Iliuchonok IR et al. Interrelations of brain hemispheres in man: Purpose, treatment of information and memory. Novosibirsk, Russia: Nauka, 1989.
41. Sheldon M, Dobson H. Reproductive challenges facing the cattle industry at the beginning of the 21st century. Reproduction. Suppl 2003; 61:1-13.
42. Rogers LJ. Lateralization in vertebrates: Its early evolution, general pattern, and development. In: Slater PJB, Rosenblatt J, Snowdon C, eds. Advances in the study of behavior. San Diego: Academic Press, 2002:31:107-162.
43. Mench J, Andrew RJ. Lateralization of a food search task in the domestic chick. Behav Neural Biol 1986; 46:107-114.
44. Robins A, Rogers LJ. Lateralized prey-catching responses in the cane toad, bufo marinus: Analysis of complex visual stimuli. AnimBehav 2004; 68:767-775.
45. Vallortigara G, Rogers LJ, Bisazza A et al. Complementary right and left hemifield use for predatory and agonistic behavior in toads. Neuroreport 1998; 9:3341-3344.
46. Andrew RJ, Tommasi L, Ford N. Motor control by vision and the evolution of cerebral lateralization. Brain Lang 2000; 73:220-235.
47. Tommasi L, Andrew RJ. The use of viewing posture to control visual processing by lateralized mechanisms. J Exp Biol 2002; 205:1451-1457.
48. Rizhova LYU, Philippova EB, Bianki VL. Characteristics of motor asymmetry in ontogenesis of the calves. Zh Vyssh Nerv Deiat Im I P Pavlova 1991; 41:1112-1118, (in Russian).
49. Manns M, Güntürkün O. Monocular deprivation alters the direction of functional and morphological asymmetries in the pigeon's (Columba livia) visual system. Behav Neurosci 1999; 113:1257-1266.
50. Prior H, Diekamp B, Güntürkün O et al. Post-hatch activity-dependent modulation of visual asymmetry formation in pigeons. Neuroreport 2004; 15:1311-1314.
51. Martin D, Weebster WG. Paw preference shifts following forced practice. Physiol Behav 1974; 13:745-748.
52. Glick SD, Carlson JN. Regional changes in brain dopamine and serotonin metabolism induced by conditioned circling in rats: Effects of water deprivation learning and individual differences in asymmetry. Brain Res 1998; 504:231-237.
53. Siegfriet B, Bures J. Conditioning compensates the neglect due to unilateral 6-OHDA lesions of substantia nigra in rats. Brain Res 1979; 167:139-155.
54. Robertson IH, North N. Spatio-motor cueing in unilateral left neglect: The role of hemispace, hand and motor activation. Neuropsychologia 1992; 30:553-563.
55. Rizhova LYu, Kokorina EP. Behavioral asymmetry is involved in regulation of autonomic processes: Left side presentation of food improves reproduction and lactation in cows. Behav Brain Res 2005; 161:75-81.
56. Babichev VN. Neurohormonal regulation of the ovarian cycle. Moscow, Russia: Medicine, 1984.
57. Cruz ME, Jamarillo LP, Dominguez R. Asymmetric ovulatory response induced by a unilateral implant of atropine in the anterior hypothalamus of cyclic rat. J Endocrinol 1989; 123:437-439.
58. Fukuda M, Yamanouchi K, Arai Y et al. Hypothlamic laterality in regulating gonadotropic function: Unilateral hypothalamic lesion and compensatory ovarian hypertrophy. Neurosci Lett 1984; 51:365-370.
59. Wittling W, Pfluger M. Neuroendocrine hemisphere asymmetries: Salivary cortisol secretion during lateralized viewing of emotion-related and neutral films. Brain Cogn 1990; 14:243-265.
60. Ahern GL, Schwartz GE. Differential lateralization for positive versus negative emotion. Neuropsychologia 1979; 17:693-697.
61. Hugdahl K. Classical conditioning and implicit learning: The right hemisphere hypothesis. In Davidson RJ, Hugdahl K, eds. Brain Asymmetry. Cambridge: MIT Press, 1995:235-267.
62. Canli T, Desmond JE, Zhao Z et al. Hemispheric asymmetry for emotional stimuli detected with fMRI. NeuroReport 1998; 9:3233-3239.
63. Henry JP. Biological basis of the stress response. NIPS 1993; 8:69-73.
64. Andrew RJ, Rogers LJ. The nature of lateralization in tetrapods. In: Rogers LJ, Andrew RJ, eds. Comparative Vertebrate Lateralization. Cambridge: University Press, 2002:94-125.
65. Denenberg VH. Hemispheric laterality in the animals and the effects of early experience. Behav Brain Sci 1981; 4:1-49.

SECTION IV
Novel Concepts in Human Studies of Asymmetrical Functions

CHAPTER 14

Asymmetry Functions and Brain Energy Homeostasis

Marina P. Chernisheva*

Abstract

Living organism is an open nonequillibrium thermodynamic system posesseing many properties which permit to evade "heat death". The analysis of these properties permits one to
imagine the general function of asymmetry in the regulation of energy homeostasis and,
in particular, a entropy production level in the organism. Here I suggest a hypothesis about the
role of autonomic nervous system laterality in specific functions of the left and right hemispheres, as well as in regulation of brain energy homeostasis. Investigations of brain asymmetries, which are associated with sex and handedness, are considered as confirmations of the
proposed hypothesis.

Introduction

While structural and functional cerebral asymmetries are essential aspects of normal
vertebrate development,[1-5] the physiological mechanisms which underlie these functional
interhemispheric asymmetries (FIHA) are poorly understood. A lot of experimental findings[6-7]
suggest that regulation of brain energy homeostasis may be one important mechanism
involved in the development of the FIHA. This hypothesis is supported by evidence showing
that asymmetry develops in tandem with increase of energy and entropy levels in animal, including human, organisms. To clarify this, some consideration of the organism's thermodynamic properties is needed.

Living Organism as a Thermodynamic System

It is a well known fact that organisms are open nonequilibrium thermodynamic systems,
which exchange with the environment substance and energy. There are some properties of
organisms, which distinguish them from other, nonliving systems. One property is a failure of
the second law of thermodynamics: "heat death" following the rise of entropy is impossible for
biosystems, which stabilize an entropy level theirselves,[8,9] but in which way or ways?

Different organism processes utilize chemical, mechanical, electrical, photic and other types
of energy. A part of the energy degrades (dissipates) in heat energy during certain biochemical
and physiological reactions. The energy may be lost by the organism by heat dispersion in
environment. The entropy is, therefore, a measure of the heat energy dispersion. This definition of entropy was formulated by R. Clausius in 1865 for an isolated (without any exchange
with environment) mechanical system. There are, though, other definitions of entropy,
e.g., for a living organism—as a measure of degree of "chaos" (or level of structural organization),[9]

*Marina P. Chernisheva—Department of General Physiology, Faculty of Biology and Soil
Sciences, Universitetskaya nab., 7/9, St. Petersburg State University, 199034, St. Petersburg,
Russia. Email: mp_chern@mail.ru

Behavioral and Morphological Asymmetries in Vertebrates, edited by Yegor B. Malashichev
and A. Wallace Deckel. ©2006 Landes Bioscience.

or as a measure of thermodynamic possibility for a process to go (or presence of energy for process realization).

In general, any influence, which rises up the nonequilibrium of an open living system, in accordance with Le Chatelie's principle, brings the system to the rise of metabolic intensity and velocity of entropy production. All organismic processes taken together produce the combined entropy of the system.[10]

Increase of information and energy input from the environment decreases the organism combined entropy.[8,9] According to Brulluan's principle, information is a factor of entropy decrease or "minus entropy" (negentropy) as it decreases uncertainty or degree of "chaos." Therefore, information has energetic nature, and it compensates the losses of energy through heat dispersion in living organisms. Information coding, decoding and transmitting between different parts of the nervous system are also connected with the rise of the level of organization (or "structurality") of energy fluxes. These processes reflect the energetic nature of information that is confirmed by energetic, electro-chemical nature of action potential as the simplest code of information in the nervous system. The next confirmation of this thesis is the use of processed information from memory as dreams or hallucinations in conditions of energetic deficiency to work with new external information, for example, during slow wave sleep (SWS), suffering or nervous system disorder in natural catastrophes, acts of terrorism, etc. This agrees with another definition of information as messages transmitting between objects (structures) by a substance and energy change.

Many specific properties of living organisms underlie the observed there decrease of the level of entropy. The most important property is the production of energy by the organism structures. For example, free energy liberates during various molecular processes, such as structural modifications and interactions between molecules, catabolic and oxidative processes, contractions of muscles and other biochemical and physiological reactions.[11] A part of this energy leads to synthesis of new molecules (ATP, GTP, etc.). It is also used in energy-capacious processes: gametogenesis, exocytose, information coding, and others. The other part of the energy dissipates in heat energy. Each stage of these processes is regulated by enzymes, which protect the system from abrupt jumps in temperature and entropy level. The heat energy disseminates both in the environment and in the organism. In the internal space the energy may be used in maintenance of enzymatic activity, somatic temperature and CNS excitability, genesis of emotions that account for need of optimal entropy level and the increase of the latter in phylogenesis. However, the number and efficiency of mechanisms, which reduce the rise of entropy, increase in evolution, as well.

Specificity of internal structures and functioning of the vertebrate organism maintain the optimal entropy level. In general, parts of physiological systems represent alterations of tubes and cavities possessing peristaltic contractions, which form pressure gradients, trends, and "structurality" of free energy flows. These properties decrease the rising of entropy level. Activities of chemo-, mechano- and thermoreceptors in visceral organs and vessels are constant generators of endogenous information (negentropy) and reinforce the effects of structural properties, which protect the organism from entropy rise.[7] Interoceptive information goes to the brain through sympathetic and parasympathetic neural pathways, and spinal tracts. Brain structures are another source of internal information, which is fixed in memory and may be used as negentropy to compensate the energy losses during heat dissipation in the organism's environment. Consequently, the rise of memory volume in phylogenesis has an energetic aspect, aside from others. It can be therefore proposed, that decrease of memory capacity and sensory functions in old humans is an important factor for raising the noncompensated energy losses (entropy) and increasing the general weakness of their organisms.

Endogenous "generators" of energy and information underlie another important property of the organism. It regulates the degree of openness of the organism as a thermodynamic system in the process of its interaction with the environment. For example, in cold conditions thermoregulation leads to decrease of such an interaction through decrease of heat-dissemination and increase of thermoisolation. Other examples of decrease of the organism-environment interactions may be the rise of introvertivity in humans (decrease of exogenous information input) during sleep, visceral pathology, or gestation. This increases the volume of endogenous information in comparison with exogenous one. This is very important for energy homeostasis especially during conditions of decrease of the organismal energetic potential or increasing energy expenditure on endogenous processes, because endogenous information is more constant, stereotypic and it processing requires less energy and time, than the processing of information from the environment. Consequently, regulation of level interaction with the environment, which is a specific property of biosystems, also contributes to control of entropy level.

Utilization of dissipative energy for organismic functions is one of the concrete mechanisms of mating the thermodynamic nonreversive processes that go with the rise of entropy, with those partially reversive processes that take place at near constant (minimal) velocity of entropy production[12] for steady state of a system.[9,13] This mating is an invention of biosystems, which underlies optimal low entropy level. There are two conditions for this mating to be fulfilled: (1) mating processes must be different in energy levels for transmission of the energy from a high energy process to the low energy process; (2) connective processes must have a common chemical and/or structural component.[12] For example, known phenomenon of oxidation and phosphorylation mating in processes of respiration and ATP synthesis. Oxidation of glucose during tissue respiration is accompanied by discharge of H_2O, CO_2 and energy to accept phosphate group by ADP at mitochondrial membranes. So, oxidation as nonreversive process is the energetic donor for phosphorylation as relative reversive process. All structural levels of animal organisms demonstrate examples of similar mating processes. Therefore, it is interesting to compare the states of stress-reaction and homeostasis on organismic level.

An organism in the state of homeostasis has a relatively constant minimal entropy level. It can be considered as analog of the steady state, which has been described for open nonequilibrium systems as "norm of chaos"[13] with possibility of reversive processes. Steady state in homeostasis characterizes by variability of basic parameters (temperature, pH, glucose concentration in plasma, temperament, etc.) at extremes of its optimal zone, which is defined by the genome. Action of exo- and endogenous stress-factors and formation of adaptive response, which requires energy, disrupt homeostasis, rising thermodynamic instability of the organism. This increases the velocity of entropy production in accordance to Le Chatelie's principle. Indeed, stress-factor as a factor of system instability leads to activation of sympatho-adrenal and visceral systems and, as a consequence, the rise of methabolism during stress-response. The rise of dissipated energy (or entropy) reflects in increase of temperature, desynchronization in EEG and negative emotions. Thus, the thermodynamic aspect of the stress-response reflects the correspondence of the latter to the state of the open system with higher instability. This state of the organism serves as an energetic donor to recover the homeostasis. The thermodynamic component associates with H. Selye's thesis on the need of stress (eustress) to maintane vital processes. Obviously, stress-response and homeostasis are thermodynamically different but mating states of the organism similar to the state of energetic donor and recipient, correspondingly (Table 1).

To sum, let us repeat briefly the properties of living organisms, which contribute to decrease the rise of common entropy. Among them are endogenous sources of energy and information (negentropy); enzymes, which regulate stage by stage the metabolic processes; mating of nonreversive and relatively reversive processes at different structural levels; organization and direction of energy fluxes by physiological systems; capability for maintenance of steady state (homeostasis) and regulation of organism's (as an open thermodynamic system) interactions with the environment.

Asymmetry as fundamental property of living organisms contributes to decrease of common entropy rise and regulation of energy homeostasis as well.[7] The following parts of this review somewhat argue in confirmation of this thesis.

Table 1. *Comparative thermodynamic characteristics of homeostasis and stress-response positions*

Characteristics of Positions	Stress-Response	Homeostasis
Equilibrium state	Maximal nonequilibrium state with rise of entropy	Near to stationary state with minimal velocity rise of entropy
Energetic potential	Donor of energy, dissipation of energy increases	Recipient of energy, heat-dissemination decreases
Correlation between reversive and nonreversive processes	Predominate nonreversive processes	Predominate relative reversive processes

General Comments on Asymmetry

On the atomic level, asymmetry is energetically more advantageous than symmetry. For example, laser-initiated asymmetrical nuclear fission of Ra^{228}, U^{234}, Fm^{256} isotopes has lower energetic threshold, and produces more total kinetic energy of the debris, than symmetrical fission.[14] Biosystems possess different asymmetries where functional asymmetries have maximal expression for thermodynamic nonequilibrium state of organism. Apparently, this explains a greater success of asymmetry investigations in different stress-models, than in experiments performed in comfort conditions.

Phenomenon of asymmetry combines unique and contradictory properties: as factor of instability, asymmetry increases the energetic expenses for destabilization of the biosystem, its energetic potential and entropy level. Simultaneously, asymmetry increases sensibility of sensory systems and contributes to the rise of conceive information (negentropy) that protects organism from the rise of entropy. Evolution of morpho-functional asymmetry at all levels of living matter shows importance of this phenomenon.

In zygote and at later developmental stages, proteins of the transforming factor beta (TGFβ) family and other factors of differentiation determine the formation of asymmetry axes and asymmetrical development of anatomical structures.[15-19] These peptides additively form asymmetrical axes, mesoderm and visceral endoderm anlagen, determine sex dimorphism, that underlie their later interactions during development. For example, bone morphogenetic proteins (BMPs) determine differentiation in gonads, lungs, skeleton, sympathetic nervous system and dopaminergic mesencephalic structures.[17-19] The processes of differentiation contribute to organization of internal space of organism, energetic flows and, consequently, protect the organism from entropy rise.

Let us say some words about the role of asymmetry axes in energy homeostasis of organism. Anterior-posterior (AP) axis is more constant in Metazoa. It is, probably, associated with rostral localization of sensory organs, which conceive and enforce input of exogenous energies and information (negentropy). Intensification of the tendency to localize sensory organs at this end of the organism in phylogenesis can be illustrated by appearance of a new homeobox gene *Arnf*, which determines development of neocortex in higher vertebrates.[20] Consequently, AP axis decreases the rise of endogenous entropy through interactions with environment.

AP axis is parallel to the Earth surface in most animals. However, it is mostly constant (in man) or transitory (during anxiety or aggression in other animals) coincide with gravitational vector. This state permits to use the gravity for acceleration of blood flow in vessels, food passing in gastro-intestinal tract, etc. It leads to the rise of metabolic velocity, energetic level and production of entropy. It is important to conceive gravitational influences by visceral mechanoreceptors. Along with proprioceptors the latter form some likeness to "peripheral vestibular organ", which supplements information that comes from cerebral vestibular organs.

Dorso-ventral (DV) axis of animal somatic asymmetry associates with constant direction of the gravitational vector perpendicular to AP axis. This stability in association with minimal variations of the Earth gravitational field leads to constancy of sensation of the somatic "map" of the organism as a factor of homeostasis. Obviously, the association of asymmetry axes with the direction of the vector of the gravitational field is conditioned by a greater stability of gravitational influences, than that of other energetic inputs, and because of the actions of gravity upon all structures of the organism.

Left-Right (LR) axis first of all is characterized by asymmetric laying of visceral organs. The laterality supposes participation of visceral receptors in estimation of direction, angles and velocity of organism rotation around the AP axis. These parameters are more variable in comparison with those associated with AP and DV axes. This might be an important factor of greater variability of LR-asymmetries. A later development of the nervous system than the visceral organs determines asymmetry of visceral sympathetic and parasympathetic nervous structures and their influences on brain asymmetry. Let us consider these influences in more details.

Asymmetry in the Autonomic Nervous System

Morphometric studies of paired sympathetic and parasympathetic structures in anuran amphibians (*Rana temporaria*) and mammals (rats, rabbits, dogs, and cats) demonstrated that peripheral sympathetic structures (number, volume and weight of ganglia and nerves) predominate on the left side, while parasympathetic—on the right side.[21-22] There are individual peculiarities in the degree of sympatho-parasympathetic asymmetry, as well. Most probably, the phenomenon underlies the known typology in humans: sympatho-, parasympatho- and mesotonics. Basic neurotransmitters determine participation of certain part of autonomic nervous system in energy homeostasis and brain functions.

Monoamines as basic transmitters of sympathetic system activate some biochemical and physiological reactions going with release of free energy and entropy rise. Investigations with the use of biochemical methods show,[23] that noradrenaline facilitates processes of glycolysis and lipolysis in peripheral tissues, but glycolysis—in brain structures. It also rises the nervous and muscle excitability, initiates contractive thermogenesis. For example, rats with high sympathetic activity[24] have higher arterial pressure in arteria carotis and arteria caudalis, higher hypertensive reaction to adrenaline injection and pulse frequency, than rats-vagotonics. In both, however, the decrease of arterial pressure after acetylcholine injection is identical. Concentration of noradrenalin in arterial probes (chronical cannule) in rats with high sympathetic activity is seven times, of adrenaline—2.5 times, corticosteron—two times as high as in animals-parasympathotonics.

Asymmetry of sympathetic structures determines laterality of its trophic effects. Greater volumes and activities of left adrenals, left lobes of glandula thyroidea and thymus, than the corresponding right structures,[25-27] confirm the influence. Functional asymmetry adds to morphological differences. In pairing test, for example, rats with right-side adrenalectomy produce more attacks, than rats with left-side adrenalectomy during first 5 min after 5 sec tail electrostimulation.[28] These morpho-functional asymmetries permit to sum up catabolic effects of steroids, thyroid hormones and noradrenergic sympathetic influences for stress-response, when energetic expenditures are maximal. Sympathetic laterality is important for asymmetry of muscles tonus and posture. So, electromyogram registration of leg extensors and flexors in rats and frogs during stimulation of ipsilateral truncus simpathicus or without it shows higher tonus of flexors on the left side, but extensors—on the right side.[21,29] This effect underlie the phenomenon of pose asymmetry, and becomes stronger due to lateralization of peptidergic spinal neurons, which secrete vasopressin.[29] Participation of noradrenaline in expression and activation of transcription factor CREB,[30] which plays an important role in long-term memory, reflects the general role of sympathetic system in control of endogenous information as negentropy.

Innervation of sensory organs, pia mater encephali, cerebral vessels (dilatation), hypophysis and epiphysis by cranial sympathetic ganglia is important for increase of metabolic activity and velocity of entropy production in the brain. Noradrenaline increases the pupil diameter and decreases sensory thresholds that lead to the rise of conceive exogenous information (negentropy) and compensation of energy losses. Investigation of sensory asymmetries in pilots and air traffic control officers shows that thresholds of olfaction, taste and tactile senses of the skin are low on their left side. For example, left nasal cavity has greater sensitivity in 71% of individuals; in 13% greater sensitivity is found on the right side, and only 6% of the population have symmetric threshold parametrs.[2] The facts coincide with the results of a morphological study, which shows that sympathetic plexus underlies submucous in the left nasal cavity, but parasympathetic plexus—in the right one.[31] Nasal receptors transmit the information by nervus olfactorius predominantly to the ipsilateral olfactory bulb.[32] Electrostimulation of the left olfactory bulb or microdose application of oxytocin to the left nasal cavity of the narcotized rat induces inspiration of longer duration, than in spontaneous respiratory cycle.[33] Stimulation on the right side leads to breaking of inspiration in favor of expiration. Therefore, sympathetic effects contribute to rise of oxidation processes, energetic potential and entropy level, contrary to parasympathetic influences.

Functions of peripheral parasympathetic structures direct the decrease in the level of oxidation processes and protect from entropy production. For example, the right nervus vagus decreases duration of inhalation, acting at the level of respiratory centre in medulla oblongata, innervates right lung lobes and liver, and controls their activities.[33,34] They, in turn, are energetically more active, than the left-sided pancreas and spleen. It is known, that general parasympathetic neurotransmitter acetylcholine acts through N- and M-receptors and generates depolarization (N) or hyper polarization (M) of cell membranes. In neurons associated with digestive system, heart, and CNS, M-receptors predominate.[29]

Some individual variability in the asymmetry of sympathetic-parasympathetic systems may be related to energetic status of organism and to the increase of entropy. That is, a predominance of sympathetic activity (sympatotonia) is coupled with the high level of energetic potential and of entropy production, while parasympathetic activity (parasympatotonia) is associated with a lower level of entropy. This resume for parasympathotonics is confirmed by deeper subcutaneous vessels and less reactive vasodilatation that decrease heat-dissemination. Consequences of the lower loss of energy are the less need for exogenous information and light energy that is reflected by smaller pupil diameter at rest because of the prevalence of parasympathetic tonus of musculus ciliaris, which constricts the iris. This contributes to greater selection of information or concentration of attention, which is characteristic for parasympathotonics. Laterality of parasympathetic influences is illustrated by more effects of right vagus electrostimulation on gastric secretion, expiration and heart contractions, than influences of the left nerve.

The autonomic nervous system is lateralized and, in turn, contributes to the asymmetrical development and activities of the CNS supraspinal structures. Conditions for this include: (1) participation of sympathetic and parasympathetic structures in the asymmetric innervation of all cerebral structures; (2) signal transmission from the peripheral sympathetic and parasympathetic structures to contralateral cerebral hemisphere via the system of lemniscus medialis, locus coeruleus and medial forebrain bundle; (3) participation of sympathetic and parasympathetic structures in regulation of brain energy homeostasis. In general, the sympathetic structures, which predominate on the peripheral level on the left side, predominantly control the right hemisphere, while the parasympathetic structures, which predominate on the peripheral level on the right side, have more influences to the left hemisphere. Support for this reasoning comes from the distribution of transmitter receptors. Adrenoreceptors have greater density in the right hemisphere, while M-cholinoreceptors have greater density in the left hemisphere. Together with distribution of glycin- and GABA receptors[29] this confirms less energetic and entropy levels in the left hemisphere, than in the right one. This receptor distribution correlates with the right > left hemisphere asymmetry of glycolysis. Levels of N-acetylaspartate and choline-containing compounds, as well as creatine/phosphocreatine metabolites using proton-MRC in human brain confirm that metabolism is higher in the right hemisphere.[6,35-36] This may coincide with vasodilatation induced by sympathetic influences. Therefore, the right hemisphere has greater level of heat dissemination, the measure of which is the entropy. It is suggestible therefore that in certain circumstances the right hemisphere may play a role of energetic donor in relation to the left hemisphere.

The results shown above indicate a possible linkage of functions of the right hemisphere with higher level of excitability and entropy production in comparison with the functions of the left hemisphere.[7] Indeed, investigations of alterations of EEG by emotional background ("tonic emotions") show, that negative emotions are associated with greater activity and heat-dissemination in the right hemisphere, while positive tonic emotions are associated in a greater extent with the left hemisphere.[37,38] Stability of negative emotional background, which is characteristic property of man in reactive depression, associates with the right neocortex as well.[2] Emotional reactions are associated with biological or social motivations, and perception of face emotional expression has more complex nature, and forms more mosaic FIHA,[38] because influences of autonomic nervous system asymmetries addict by exogenous information for energetic maintenance of emotions.

To the lower energetic potential of the left hemisphere point not only its association to the positive emotions, greater content of inhibitory mediators (GAMK, glycine), but also its main function—speech and its cognition, and interrelated consequent-logical processing of information. Indeed, speech during visual and auditory perception is a more simple and stereotyped signal in comparison to other sensory inputs of the environment; it needs less energy for its processing.

Important factors of FIHA, as well as energy homeostasis of brain and organism, are sex and handedness.

Hemispheric Asymmetry and Sex

Sex hormones are well-known for direct influence on the development of sexual dimorphism in the liver, brain and other structures from as early as 7-12 weeks of gestation in men. Investigations of sexually dimorphic structures in CNS by morphological and immuno-histochemical methods show asymmetry in structural sizes, neuron density and affinities to sex steroids. There are asymmetries of spinal nuclei of nervi ishio- and bulbocavernosi,[39] with axonal innervation of the external genitalia, adrenergic groups of neurons in medulla oblongata,[40] which accept sex steroids and are included in respiratory and vasomotor centers. This explains gender specificity of respiration and cardiac activity parameters and their alterations in different stages of estrous cycle. The nucleus of sex dimorphism in medial hypothalamus, which neurons bind sex steroids and secret gonadoliberin is greater and more asymmetric in male rats.[41] In male and female rats the asymmetry of amygdaloidal nuclei was shown, that is in difference in size and localization of zones, which bind progesterone, estrogens, and androgens.[42] This determines gender specificity of olfactory sensitivity, sex and other forms of behavior. Asymmetry of binding of sex steroids by the neurons in the bed nucleus of stria terminalis, connecting hypothalamus with the amygdala,[42] indicate a hierarchy of manifestation of asymmetry in cerebral structures. In caput of nucleus caudatus (ventral and dorsal parts), which takes part in motor and mental functions, density of neurons and neurons/satellite glia proportion is higher in women, than in men, and greater in the right, than in the left structure.[4] Laterality of peripheral reproductive organs make specific corrections in FIHA: the power of EEG rhythms in women with left and right attaching of placenta show greater effect in contralateral hemisphere.[43] The steroid influences on cortex asymmetry were also described.[44]

Higher energetic potential in men compared to women leads to prolonged gametogenesis on the relative constant level. This energetic specificity of men's organism is genetically determined and reflected by greater physiological and biochemical parameters such as vital capacity of lung, force of cardiac contraction, hemoglobin content, skeletal muscle mass, level of oxidative metabolism, catabolic activity of liver, somatic temperature. Together with greater metabolic effects of androgens, than estrogens, this leads to intense thermogenesis and considerable increasing of entropy as a measure of heat energy dissemination. It is known, that dominance of the left hemisphere is characteristic property of the male brain.[1-3] It may be, therefore, assumed that increasing dominance of the left hemisphere, decreasing of entropy level, as well as the relative stability of the FIHA type are an adaptations of the male organism to these rises in common entropy. This is confirmed by results[45] of psychological testing (tests TOPOS, BTSA, Kattle 17 LF) of men with pathology (bronchial asthma) in groups of 6-8, 11-14, 20-35 and 45-60 year old individuals, which showed a rise of stability of left-hemispheric dominance during increase of illness duration (from 2 to 23 years). During processing of verbal information, the man's left hemisphere activates diffusively and performs significant inhibition of the right hemisphere, whereas in the women's left brain only two zones of activation have been localized, and their inhibition of the right hemisphere is weak.[46]

In contrast to men, in women the right hemisphere is dominant in majority of functions. This may have an evolutionary advantage by altering maternal behavior and by causing the mother to seek environments, which expose the child to low levels of threat. Because of the preponderance of right hemisphere dominance and the associated increase in right-hemisphere excitability, women are able to conceive larger amounts of information, that requires an additional expenses of energy and guides to additional entropy rise. The latter is supported by the fact of greater emotional responsiveness and participation of the right hemisphere in mediating prosody.[1-3] In processing of verbal information women's left hemisphere has less inhibiting influence to the right one, than in men.[46] This suggests that energetic adaptation to right-hemisphere dominance in women are: ability to conceive large amounts of information by subconscious (without words), and a greater lability of excitation and inhibition processes, which are determined by greater than in men's brain number of fibers in posterior part of corpus callosum.[47] This promotes generalization of sensory information, which has inputs in occipital and parietal lobes of both hemispheres, and contributes to use of left-hemisphere functions for better organization of information and, consequently, protection of entropy rise. These facts are confirmed by a state "without dominances," which is a characteristic state for women's brain during fatigue, i.e., losses of energy and entropy rise. Therefore, the state "without dominance" can be considered as an adaptive regime of homeostatic regulation of the entropy level by the women's brain. Recent studies showed that girls who adopt the traditionally male habits of smoking and alcoholism have clear dominance of the left hemisphere (on leading range of vision) and associated decrease in lability. Additionally, the threat of miscarriage in women during the six to seven months of pregnancy causes a switch from right-hemisphere dominance in alpha-1, alpha-2 rhythms in EEG to left hemisphere pattern of dominance.[48] Consequently, the right-, left-hemisphere dominance or the state "without dominance" in women's brain reflect the regimes of dynamic asymmetry, that are adaptive to regulation of entropy level. Thus, the requirements of energy homeostasis determine the gender properties of FIHA.

Interhemispheric Asymmetry and Motor Activity

Among the clearest examples of motor FIHA are lateralized motor reactions. Individuals may have same side, or opposite side, preferences for arm and leg use.[2] Hands as "sensory organs" are localized closer to the head and make a greater contribution to the active "collection" of exogenous information (negentropy), than legs, hence reinforcing the AP asymmetry. In man, arms' movements are rather diverse compared with more stereotyped leg motor reactions, and it is likely that cerebral sensory organs take a greater role in the control of their movements.

In terrestrial vertebrates, genetically determined handedness may vary on species or individual level.[1,5] Determinants of handedness are influenced by a variety of variables, including the genetically determined degree of animal asymmetry, the ecological niche peculiarities,[49] the frequency of vertical vs. horizontal motor activity, the effects of gravity on posture, and energetically favorable motor responses. In humans, the genetic predisposition for right-handedness suggests that a right-handed man has greater chance for survival. In accordance to the hypothesis described here, the dominance in right-handers of the left hemisphere, which is presumably "parasympathetic" by informational inputs and "negentropic" by functions, in the control of different motor reactions of the leading (and therefore pro-entropic) right hand makes the movement pattern better and protect from the entropy increase during movement and manipulation.

This confirmed by greater stereotypy (automatism), discreteness and the amplitude of the right hand movements.[2,50] In stress conditions, the right hand increases velocity, coordination and accuracy of movements, and the latency of reaction what is connected to the greater in comparison with the left hand sensitivity to movements and more pronounced verbal function of the left hemisphere.[51]

In contrast, in left-handers summation of the heat dissemination of the leading hand and the right ("sympathetic") hemisphere increase the entropy production. It follows in right-handed men by the lower temperature on the back of the hand and lower cutaneous potentials on the right hand,[52-53] than on the left one. Evidently, the higher temperature of the left hand determines its higher sensitivity to multimodal influences. This forms lower thresholds of sensitivity to tactile inputs, pain and vibration in the left, than in the right hand.[2]

Greater excitability, facilitation of the generalized excitation, and increased duration of track processes reflect increased entropy production in left-handers and are the reasons of higher probability of nerve and visceral pathologies.[54] It is characteristic for left-handed and ambidextrous men to show compensatory increase in lability of processes of excitation transmission from one hemispheres to the other, which is possible because of increase in the number of fibers in the rostral part of corpus callosum, which connect the frontal cortex of the two hemispheres.[55,56] This leads to decrease of FIHA coefficients in motor and emotional reactions,[2] but increases the probability of unusual combinations of functions and appearance of extraordinary talents in left-handers.

Some authors consider ambidexterity as latent left-handedness. But genetically determined ambidexterity is often coupled with the absence of expressed FIHA not only for motor function, but for some other functions as well. Absence of the interhemispheric asymmetry may correspond to the lack of asymmetries of the autonomic nervous system that is normal in man with mesotonia. Ambidexterity in relearned left-handers have an enhanced risk of neuroses and psychosomatic disturbances.[2,54] This assumes that the phenomenon is a consequence of left-hemisphere dominance absence in control of entropy level for motor reactions by the leading hand, which genetic right-handers have.

Functional interhemispheric asymmetry is apparently one of the most powerful system mechanisms of controlling the level of entropy production and regulation of energy homeostasis. A basic endogenous factor, which plays a considerable role in formation of brain asymmetry, is the asymmetry of sympathetic and parasympathetic influences.

References

1. Bianki VL. Individual and species asymmetries in animals. Zhournal Vysshei Nervnoi Dejatelnosti 1979; 29:295-304, (In Russian).
2. Dobrohotova TA, Bragina NN. Methodological role of symmetric principle in investigations of human functional organization. In: Bogolepova NN, Fokin BF, eds. Functional Interhemisphere Asymmetry. Moscow: Scentific Centre, 2004:15-46, (In Russian).
3. Andrew RJ. Memory formation and brain lateralization. In: Rogers LJ, Andrew RJ, eds. Comparative Vertebrate Lateralization. Cambridge: Cambridge University Press, 2002:582-641.
4. Bogolepova NN, Malofeeva NS, Orgehovskaja TV et al. Cytoarchitectonic asymmetry of neocortex areas and nucleus caudatus in humans. In: Bogolepova IN, Fokin BF, eds. Functional Interhemisphere Asymmetry. Moscow: Scentific Centre, 2004:191-205, (In Russian).
5. Rogers LJ. Lateralised brain function in anurans: Comparison to lateralization in other vertebrates. Laterality 2002; 7:219-240.
6. Braun CMJ, Boulanger Y, Labelle M et al. Brain metabolic differences as a function of hemishere, writing hand preference, and gender. Laterality 2002; 7:97-113.
7. Chernisheva MP. Interhemisphere asymmetry and entropy. In: Dubrovsky, ed. Functional Interhemisphere Brain Asymmetry. Moscow: Institute of Brain RAN, 2003:156-164, (In Russian).

8. Schrödinger E. What is life? Physical aspect of living cell. Cambridge: Cambridge University Press, 1944, (Cited after Russian edition: Moscow: R and D Dynamics. 2002).

9. Prigogine I, Stengers I. Order out of Chaos. Man's new dialog with nature. Toronto: Bantom Books, 1984.

10. Levich AP. Time and entropy. Vestnik of Russian Humanitarian Research Foundation, 2002:110-115, (In Russian).

11. Rubin BA. Biophysics. Vol. 1. Moscow: Nauka 2004, (In Russian).

12. Opritov VA. Entropy of biosystems. Soros Education J 2000; 6(3):123-134, (In Russian).

13. Klimontovich YUL. Relative ordering criteria in open systems. Physics-Uspekhi 1996; 39(11):1169-1175, (In Russian).

14. Møller P, Madland DG, Sierk AJ et al. Nuclear fission modes a fragment mass asymmetries in a five-dimentional deformation space. Nature 2001; 409:785-790.

15. Yost HJ. Left-right development from embryos to brains. Dev Gen 1998; 23:159-163.

16. Wood WB. Left-right asymmetry in animal development. Ann Rev Cell Devel Biol 1997; 13:53-84.

17. Wagner DS, Mulling MC. Modulation of BMP activing in dorsal-ventral pattern formation by the Chordin and Ogon antagonists. Dev Biol 2002; 245:109-123.

18. von Bubnoff A, Cho KWY. Intracellular BMP signaling regulation in vertebrates: Pathway or network? Dev Biol 2001; 239:1-14.

19. Lane AH, Donahoe PK. New insights into Mullerian Inhibiting Substance and its mechanisms of action. J Endocrin 1998; 158:1-6.

20. Aleksandrova EM, Zaraiski AG. Molecular mechanism of early neurogenesis in vertebrates. Mol Biol 2000; 34:496-507, (In Russian).

21. Osipova NC. On question about phenomens of functional asymmetry in autonomic nervous system. Abstr. V Conference of ANS Physiology in Memory of Orbedi LA. Erevan: 1982:67, (In Russian).

22. Pushkarev UP, Gerasimov AP. Phenomen of asymmetry in visceral systems. Ugolev AM. Abstr. Int. Conf. dedicated to 75-year. St. Petersburg: 2001:141, (In Russian).

23. Eschenko ND. Energetic methabolism in brain. In: Ashmarin IP, ed. Biochemistry of Brain. St. Petersburg: St. Petersburg University Press, 1999:124-168, (In Russian).

24. Fedorov VI, Cherkesova OP. Influence of stable acetylcholine analog and ACE activity in lungs, gl. renalis and plasma in rats with high sympathetic activity. IM Sechenov Physiological Journal 1997; 83:76-83, (In Russian).

25. Gontova IA, Abramov VV, Kozlov VA. Laterality of thymic lobes and immune response in mouse (CBA × C57BL/6F1). Immunology 2000; 30-32, (In Russian).

26. Perelmuter VM. Functional asymmetry of thymicoadrenal system and adrenal zona fasciculata during homotransplantation of right and left thymus parts. Bull Exp Biol Med 1998; 124:577-579, (In Russian).

27. Nozdrachev AD, Kovalenko RI, Chernisheva MP. The functional asymmetry of adrenal glands induced by pineal peptides and oxytocin in stress. Europ J Physiol 1995; 430:137-139.

28. Kovalenko RI, Zymina OA, Chernisheva MP et al. Effects of intranasal administration of pineal-gland peptides on aggressive-defensive behavior of unilaterally adrenalectomized rats. Doklady Akadi Sci 2000; 370:17-20, (In Russian).

29. Vartanjan GA, Klementjev BI. Problem of chemical brain asymmetry. Human Physiology 1988; 14:297-306, (In Russian).

30. Simmoneaux V, Ribelayga CH. Generation of melatonin endocrine message in Mammals: A review of the complex regulation of melatonin synthesis by norepinephrine, peptides and other pineal transmitters. Pharm Rev 2003; 55:325-395.

31. Heine O, Galaburda AM. Olfactory asymmetry in the rat brain. Exp Neurol 1986; 91:392-398.

32. Voronkov GS. Neuromorphology of olfactory pathways in Mammalia. IP Pavlov J Higher Nervous Activity 1994; 30:432-453, (In Russian).

33. Chernisheva MP, Osipova NS. The vagus influence on olfactory visceral control. Canada: III Congress IBRO, 1991:189.

34. Grelot L. Responses in inspiratory neurones of the dorsal group to stimulation of expiratory muscle and vagal afferents. Brain Res 1999; 507:281-288.

35. Gur RC, Mozley LH, Mozley PD et al. Sex differences in regional cerebral glucose metabolism during a resting state. Science 1995; 267:528-531.

36. Powels PJ, Frahm J. Regional metabolite concentrations in human brain as determined by quantitative localized proton MRS. Magnetic Resonance Med 1998; 39:53-60.

37. Davidson RJ. Anterior cerebral asymmetry and the nature of emotion. Brain Res 1992; 236:125-151.

38. Rusalova MN. Functional asymmetry: Emotion and activity. In: Bogolepova IN, Fokin BF, eds. Functional Interhemisphere Asymmetry. Moscow: Scentific Centre, 2004:322-348, (In Russian).

39. Breedlove SM, Arnold AP. Sexually dimorphic nucleus in the rat lumbar spinal cord. Brain Res 1981; 225:297-307.

40. Haywood SA, Sinonian SHX, van der Beek EM et al. Fluctuating estrogen and progesterone receptor expression in brain stem norepinephrine neurons through the rat estrous cycle. Endocrinol 1999; 140:3255-3263.

41. Gorski RA, Gordon JH, Shryne JE et al. Evidence for a morphological sex difference within the medial preoptic area of the rat brain. Brain Res 1978; 148:333-346.

42. Akmaev IG, Kallimulina LB. Amygdaloid complex: Functional morphology and neuroendocrinology. Moscow: Nauka, 1993, (In Russian).

43. Orlov VI, Chernositov AV, Sagamonova KU et al. Interhemisphere asymmetry of brain in systematic organization of women's reproduction. In: Bogolepova IN, Fokin BF, eds. Functional Interhemisphere Asymmetry. Moscow: Scentific Centre, 2004:411-443, (In Russian).

44. Lewis DW, Damond MC. The influence of gonadal steroid on the asymmetry of the cerebral cortex. In: Davidson RU, Hughal K, eds. Brain Asymmetry. Cambridge: MIT Press, 1998:456-601.

45. Gorskaja EA. Psychological specifities of children and men with bronchial asthma sickness. PhD Thesis (Abstract). St.-Petersburg Pedag: St. Univ., 2005:1-16 (In Russian).

46. Volf NV, Rasumnikova OM. Sex dimorphysm of brain functional organization of verbal information processing. In: Bogolepova IN, Fokin BF, eds. Functional Interhemisphere Asymmetry. Moscow: Scentific Centre, 2004:387-410, (In Russian).

47. De Lacosta-Utamsing MC, Holloway RL. Sexual dimorphism in the human corpus callosum. Science 1982; 216:1431-1432.

48. Smirnov AG, Batuev AC, Vorobjev SU. Specifity of EEG in women with pathologies in gestation. Human Physiology 2002; 28:56-66, (In Russian).

49. Malashichev YB, Wassersug RJ. Left and right in amphibian world: Which way to develop and where to turn? BioEssays 2004; 26:512-522.

50. Kimura D, Humphres CA. A comparison of left- and right-arms movements during speaking. Neuropsychol 1981; 19:807-812.

51. Ananjev BA. Bilateral regulation as mechanism of behavior. Voprosi Psychologii 1963:89-98, (In Russian).

52. Matojan RA. Laterality of tactile conceivation in left- and right-handed men. Human Physiology 1998; 24:131-133, (In Russian).

53. Kabanova NP, Chepurnova NV, Chepurnov CA. Modifications of cutaneous electroconductivity by different doses of intranasal thyroliberin. Medicothechnical questions of reflexotherapy, physiology and eviromental control. Tver: Tver State Univ Press, 1992:33-38, (In Russian).

54. Mayer AR, Kosson DS. Handedness and psychopathy. Neuropsychiatry, Neuropsychology and Behavioral Neurology 2000; 13:233-238.

55. Witelson SF. The brain connection: The corpus callosum is lager in left handers. Science 1991; 229:665-668.

56. Cowell SE, Denenberg VH. Development of laterality and the role of corpus callosum in rodent and humans. In: Rogers LJ, Andrew RJ, eds. Comparative Vertebrate Lateralization. Cambridge: Cambridge Univ Press, 2002:574-305.

Index

Laterality

Asymmetries of Body, Brain and Cognition

This journal publishes high quality research on all aspects of lateralisation in humans and non-human species. *Laterality*'s principal interest is in the psychological, behavioural and neurological correlates of lateralisation. The editors will also consider accessible papers from any discipline which can illuminate the general problems of the evolution of biological and neural asymmetry, papers on the cultural, linguistic, artistic and social consequences of lateral asymmetry, and papers on its historical origins and development. The interests of workers in laterality are typically broad. Submission of interdisciplinary work, either empirical or theoretical, or concerned with problems of measurement or statistical analysis, is therefore also encouraged. The journal publishes special issues on particular topics. The first issue was published in March 1996, and six issues are now published every year.

The editors encourage the submission of dissenting opinions and comments that directly relate to papers that have been published in *Laterality*. The editors reserve the right to terminate an interchange after a comment and response to the comment. Comments and responses to comments will be subject to the normal review process.

VOLUME 11 (2006) 6 ISSUES PER YEAR
PRINT ISSN: 1357-650X • ONLINE ISSN: 1464-0678
INSTITUTIONAL RATE: £349 / $577
PERSONAL RATE: £164 / $271

For more about this journal, including full instructions for authors, a free online sample copy, previous contents, and Preview (forthcoming articles online) visit

www.psypress.com/laterality

For more titles and resources in these areas:
www.cognitivepsychologyarena.com
www.neuropsychologyarena.com

Taylor & Francis Group